Dispersing Additives for Plastics Processing

From Formulation to Field: Applications, Troubleshooting & Case Studies

A Comprehensive Resource for Formulators, Producers, Users and Consultants

About the Book

"Dispersing Additives for Plastics Processing", Volume 1 in *"The Plastics Additives and Modifiers"* Series" is not a theory manual. It's a field guide, written for compounders, additives and masterbatch formulators, extrusion engineers, and technical managers who need answers that hold up when the line is running, the customer is waiting, and the batch can't afford to fail.

This book distills over two decades of real-world formulation work into a structured, actionable reference. You'll find no marketing fluff, no hypothetical "ideal-case" scenarios, and no vague recommendations. Instead, you'll get:

Ten fully documented formulations, from PE pigment masterbatches to bio-based PLA systems, complete with dosages, processing windows, KPIs, safety notes, and optimization tips, all pressure-tested on production equipment.

A troubleshooting framework that turns visible defects, specks, haze, plate-out, brittleness, into root-cause diagnoses, backed by field case studies and a cause/effect matrix.

Decision tools, not just data: the 5-Question Additive Selector, KPI Threshold Tables, and Additive Incompatibility Matrix help you eliminate trial-and-error before the first pellet hits the extruder.

Compliance made practical: clear guidance on FDA, REACH, and recyclability, not as regulatory text, but as formulation constraints you can design around from day one.

Future-ready insights on AI-assisted formulation, smart (responsive) additives, and regional regulatory shifts, so you're not just solving today's problems, but anticipating tomorrows.

Whether you're scaling a recycled-content compound in the Gulf, qualifying a medical-grade masterbatch for EU audit, or troubleshooting die build-up at 2 a.m., this book gives you the why behind the what, so you can adapt, validate, and succeed, every time.

Volume 1 focuses on foundations: selection, formulation, validation, and failure analysis.

Because in compounding, the goal isn't just to mix.
It's to anchor, molecule by molecule, batch after batch.

American Industrial Technologies LLC.

About the Series

"The Plastics Additives and Modifiers Library" is a comprehensive technical reference series dedicated to the science, formulation, and industrial applications of plastics additives. Each volume in the series explores a specialized category of additives or modifiers, ranging from dispersants, stabilizers, plasticizers, compatibilizers, and coupling agents to performance enhancers, fillers, and colorants, presented with a focus on real-world use, formulation design, and process integration.

Written for additives formulators, compounders, product developers, and process engineers, the series bridges the gap between chemistry science, industry technology, and production-scale practice. It distills decades of industrial best practices, accumulated know-how, and formulation insight into accessible, technically rigorous guides that support problem-solving and innovation across the polymer industry.

Whether you're developing common additives and modifiers, additives-based masterbatches, next-generation sustainable compounds, troubleshooting complex material behavior, or optimizing additive packages for performance and cost, "*The Plastics Additives and Modifiers Library*" provides the scientific and practical knowledge bases and technical depth needed to move from concept to successful formulation.

Dispersing Additives for Plastics Processing

From Formulation to Field: Applications, Troubleshooting & Case Studies

A Comprehensive Resource for Formulators, Producers, Users and Consultants

1st Edition

Volume 1

Copyright

Dispersing Additives for Plastics Processing: From Formulation to Field Applications, Troubleshooting & Case Studies
Author: Mostafa Ebrahim, PhD
Published by: American Industrial Technologies LLC.

Disclaimer

The field of plastics additives and modifiers is continually advancing. As new research, industry standards, and practical experiences emerge, formulations, application techniques, and troubleshooting methods may evolve.

The content of this book reflects the author's knowledge and perspectives at the time of writing and does not necessarily represent the views of the publisher or editors. Neither the author nor the publisher guarantees the accuracy, completeness, or suitability of the information provided for any specific purpose.

The reader assumes all responsibility for the proper application, testing, and verification of any procedures, formulations, or recommendations outlined in this book. The author and publisher disclaim any liability for damages, including but not limited to structural failure, material incompatibility, or financial loss, arising from the use or misuse of the information contained herein.

Professionals and practitioners are advised to consult current industry standards, local regulations, and qualified experts before implementing any techniques or formulations in real-world applications.

Trademark and Brand Disclaimer

All trademarks, service marks, trade names, trade dress, logos, and other brand identifiers featured in this publication are the property of their respective owners. Their inclusion is solely for identification and illustrative purposes and does not imply any endorsement or affiliation with the book's author or publisher. This publication does not intend to infringe upon the rights of any trademark holder.

Intended Use Statement

This book is a hands-on reference for compounders, formulators, process engineers, and technical managers working with thermoplastics and masterbatches. It's built on real-world experience, not theory, and reflects proven practices at the time of writing. It does not replace your judgment, testing, regulatory checks, or adherence to standards like FDA, REACH, ASTM, or ISO.

In short: Let it guide your decisions, but always test, validate, document, and then scale.

Complimentary Technical Advice

As a thank-you to registered readers, I offer free technical support via email to help with questions or challenges related to plastics additives and modifiers formulation and application. Whether you need clarification, product guidance, or best practice recommendations, I'm available to assist.

This service is provided for the contents of the purchased book and is limited to 2 hours per reader and must be requested within one month of purchasing a new copy from an authorized source. Please use the email address associated with your purchase to ensure prompt assistance.

Please note: This advisory service is provided voluntarily and does not constitute a contractual obligation. It reflects my commitment to supporting professionals in the field and promoting knowledge sharing across the industry.

I look forward to engaging with readers and contributing to their success through this opportunity.

Dedication

To my beloved father, whose steadfast guidance and unwavering work ethic have been a constant source of inspiration. Thank you for teaching me the true value of perseverance and the relentless pursuit of excellence.

To my caring mother, whose unconditional love and nurturing spirit have profoundly shaped me into the person I am today. Your endless encouragement has empowered me to boldly chase my dreams and passions.

To my wonderful wife, Monica, whose unwavering support and love have been my anchor. Your belief in me and our shared dreams inspire me every day.

And to my darling daughters, Fatima and Mona, whose boundless enthusiasm and insatiable curiosity remind me daily of the profound importance of wonder, discovery, and building a better future. This work is a testament to the world I hope to create for you.

This book is a humble offering to the four pillars of my life, may it serve as a small expression of my deepest, most heartfelt love and gratitude.

Acknowledgments

The knowledge in this book didn't come from a lab bench alone, it came from the plant floor: from extruders that wouldn't stop clogging, dies that gummed up at midnight, and batches that passed visual inspection but failed in the field. What began as troubleshooting notes, formulation logs, and hard-won lessons from failed trials has grown, through years of collaboration, with compounders, masterbatch producers, and process engineers across three continents.

I'm deeply grateful to the technical teams and production managers who opened their facilities, shared real-time data, and let me stand beside the line while we diagnosed specks, plate-out, and migration, not in theory, but in real time. Your honesty, pragmatism, and refusal to accept "good enough" shaped every chapter.

Special thanks to my wife, Monica, and our daughters, Mona and Fatima, for your patience during early mornings and late nights, and for tolerating a kitchen table that was more spreadsheet than breakfast nook. This work wouldn't exist without your steadfast support.

I also thank my colleagues at American Industrial Technologies LLC and the broader polymer community, especially the many anonymous reviewers who challenged assumptions, flagged oversights, and insisted on clarity over cleverness. Your rigor ensured this book stays grounded in what works, not just what sounds right.

Finally, to the next generation of formulators: may this book help you skip the trial-and-error, avoid the costly mistakes, and get to first-pass success, faster, safer, and with more confidence.

This isn't a solo effort. It's a field manual, written by the industry, for the industry.

Dr. Mostafa Ebrahim
Additives Formulator and Consultant
American Industrial Technologies LLC.

Forward

When I first started in compounding, I thought dispersion was about mixing harder, hotter, and faster. I ran twin-screw extruders at redline, overloaded feeders with pigment, and added more EBS, just in case. And yet, the specks remained. The plates built up. The batches failed at the customer.

It took years, and many costly failures, to realize: dispersion isn't about force. It's about fit. The right additive, in the right amount, at the right time, in the right polymer. Not more. Not louder. Just right.

That's the insight this book delivers, not as theory, but as hard-won practice.

Dr. Mostafa Ebrahim has done something rare: he's taken the tacit knowledge of seasoned compounders, the kind that lives in shift logs, maintenance notes, and 3 a.m. troubleshooting calls, and structured it into a clear, actionable reference. Every formulation, KPI, and case study here has been pressure-tested on real lines, with real scrap budgets and real customer deadlines.

You won't find marketing claims here. No "universal" solutions. No vague promises. What you will find are the precise boundaries of what works, and why things fail when you cross them. You'll learn why EBS seems to solve a problem today but guarantees one tomorrow. Why a 0.5% change in dispersant can mean the difference between a 48-hour run and a 4-hour shutdown. And why sustainability isn't a checkbox, but a chemical commitment that starts at the molecular interface.

This book won't replace your experience.

But it will help you earn that experience faster, and avoid the mistakes that waste time, material, and trust.

I keep a copy in our lab, dog-eared and annotated.

You'll want one on your desk, too.

Adel Bagati, Eng.
Plastic Compounding Consultant
American Industrial Technologies LLC.

About the Author

Dr. Mostafa Ebrahim is a polymer and additives expert, technical writer, and entrepreneur with over two decades of hands-on experience in plastics compounding, masterbatch formulation, and sustainable additives development. His work sits at the intersection of applied science and real-world production, where theory meets torque, and lab data must survive the extruder.

He is the author of the Concrete Chemical Admixtures series, a body of work that, like this book, bridges deep technical science with real-world practice, transforming complex material behavior into reliable, scalable, and performance-driven solutions.

Mostafa is a co-founder and the lead authority on additives formulation and applications at American Industrial Technologies LLC, where he partners with compounders, packaging innovators, and circular-economy startups to solve dispersion challenges in recycled, bio-based, and high-performance polymer systems. His formulations are used in automotive interiors, medical devices, food packaging, and infrastructure-grade pipe, always with the same goal: fewer specks, less scrap, more uptime.

He continues to work directly on the plant floor, running trials, reviewing filter packs, and training technicians, not from a desk, but from the control panel, where the real decisions happen.

When he's not troubleshooting die build-up or validating KPIs, he's at home with his wife, Monica, and their daughters, Mona and Fatima, debating which of his Karak Tea formulations to brew.

American Industrial Technologies LLC.

Preface

This book was born not in a research lab, but on the plant floor.

It began with a speck, just one, on a black automotive trim part that passed visual inspection but failed the customer's 10× magnification audit. It grew from a film line shut down at 3 a.m. because the die wouldn't stay clean. It was shaped by the quiet frustration of a compounder who kept adding more EBS, hoping this batch would finally run cleanly.

Over two decades, I've watched brilliant engineers and skilled technicians solve the same dispersion problems again and again, because the knowledge lived in notebooks, shift logs, and verbal handovers, not in a shared, structured reference. Too often, we mistook symptoms for causes: blaming the pigment when the dispersant was incompatible, blaming the extruder when the additive was degrading, blaming the operator when the formulation had no margin for real-world variation.

This book is an attempt to change that.

It is not a theoretical treatise. It does not prioritize chemical elegance over operational reality. Instead, it distills field-tested insights, from HDPE pipe extrusion in Texas to medical-grade compounding in Bavaria, from recycled-content startups in Dubai to high-gloss film lines in Thailand, into a clear, actionable framework.

This is the book I wish I'd had when I started. The one that doesn't just tell you what to do, but why it works, when it breaks, and how to prove it before you scale.

To my colleagues in compounding: may it help you spend less time troubleshooting, and more time innovating.

Mostafa Ebrahim
Additives Formulator and Consultant
American Industrial Technologies LLC.

Disclosure: Acknowledging Collaborative Technology

This book has leveraged advanced writing technology as a supportive tool. While all key concepts, technical analyses, and original insights remain the author's own work, artificial intelligence assisted by suggesting relevant research, verifying facts, improving grammar, and enhancing readability through structural refinements.

This technological collaboration was designed to complement human expertise, not replace it. The author maintains complete oversight and responsibility for all technical accuracy, professional judgments, and the work's overall integrity. Every recommendation and conclusion has been carefully reviewed to meet industry standards for plastics technology.

This project represents an innovative approach in technical publishing, where AI augmentation helps subject-matter experts produce more precise, thorough, and accessible content. As you engage with this material, we hope you'll appreciate how thoughtfully applied technology can enhance - while never supplanting - expert knowledge and rigorous scholarship.

Declaration of Competing Interest

I hereby declare that there are no known competing financial interests, professional affiliations, or personal relationships that could be perceived as influencing this publication, content, or recommendations presented in this work. No financial support or compensation was received from any organization that might have an interest in the outcomes of this publication.

Mostafa Ebrahimi

1. Introduction

Dispersion isn't just a mixing step; it's the foundation of quality. When solid additives like pigments, fillers, or fibers aren't properly wetted and dispersed, they form agglomerates that lead to specks, weak mechanical properties, and production headaches. This chapter explains why dispersion matters, what happens when it fails, and how the right additives bridge the gap between raw materials and reliable performance.

1.1 Scope Of the Book

To deliver a practical, no-fluff, field-tested reference that helps professionals choose, formulate, apply, and validate dispersing and wetting additives, without bias, without hype, and without compromise. just actionable insight that works on the plant floor.

1.2 The Need for Dispersing and Wetting Additives

When processing plastics, ensuring a uniform distribution of colorants, fillers, flame retardants, and reinforcing materials is important, not only from an aesthetic viewpoint, but also with respect to their mechanical properties. If such solid additives aren't properly wetted and dispersed, they usually agglomerate. This is due to their high forces of attraction, such as in the case of van der Waals attractive forces, causing shortcomings such as the formation of specks, streaks, inconsistent or poor color strength, filter clogging or screen pack blockage, or impact/tensile strength.

These factors are even more significant when high filling levels of inorganic materials (e.g., $CaCO_3$, talc) or fine organic colors are added to non-polar polymers such as polyethylene or polypropylene, in which cases the natural affinity between the molten polymer and the additive surfaces is low. The addition of dispersing and wetting agents bridges such gaps by reducing the tension between the interfaces, helping separate the particles, and ensuring their proper incorporation.

1.3 Definition

Dispersing and wetting agents are high-performance processing agents added specifically to facilitate the processing of solid particulates into a polymer matrix. These agents work by binding to the surfaces of solid particulates, lowering the interfacial tension between the processing agent and the molten polymer, and inhibiting the tendency of the particulates to re-agglomerate during processing. 'Wetting' essentially describes the displacement of air or liquid surface coating on solid particulates with molten polymer, while 'dispersion' describes the process of reducing solid particulate agglomerates to primary particles.

1.4 Common Misconceptions

Several misconceptions related to the dispersing and wetting agents used in everyday compounding and processing might result in suboptimal outcomes or unnecessary expenditure. Dispersing agents enable the mixing of high amounts of powdered ingredients by reducing the viscosity.

All lubricants are dispersants

Although certain lubricants (EBS, PE Wax) have some secondary dispersion properties, true dispersants have been designed to specifically adsorb on the surface of particles, contributing to steric or electrostatic stabilizations. Using lubricants with general-purpose properties is often inadequate in resolving agglomeration in high-dispersion products such as high-pigmented masterbatches or compounds containing minerals.

Higher additive dosage always improves dispersion

Excess dosage could result in plate-out on the dies and rolls, surface blooming, lower mechanical properties, or even affects certain operations such as printing or painting. Dispersion is all about optimal dosage, not maximum dosage.

If the mix looks uniform, dispersion is good

Visual homogeneity does not always prove to be ideal. There might be microscopic agglomerates, even if the visually detected homogeneity is satisfactory, which work as points of stress and therefore cause the impact strengths and sometimes the filter clogging during extrusion to be affected.

One dispersant works for all polymers

Dispersants should be of the same polarity and processing window as the base polymer. The dispersant, made from surfactants, which works well in PET, might be unsuitable, or even damaging, when used in polypropylene.

Dispersion is only a compounding issue

"A poor initial dispersion cannot be 'fixed' in injection molding or in film blowing." The initial melt mixing is when the quality of dispersion is mostly established, while other processing steps just uncover the outcome. By clearing up such misconceptions right from the start, the need for trial-and-error problem solving in relation to their treatment can be avoided,

1.5 Scope

This guide focuses on the practical application of dispersing and wetting additives in thermoplastic compounding and processing. It is intended for plastics compounders, masterbatch producers, extrusion engineers, and formulation specialists working with polyolefins, engineering plastics, PVC, and filled polymer systems.

The content covers:

- How dispersion and wetting challenges manifest in real-world production,
- The functional mechanisms and chemistry of common additive types,
- Selection criteria based on polymer matrix, filler/pigment type, and processing conditions,
- Dosage guidelines, performance evaluation methods, and compatibility considerations,

- Regulatory constraints and illustrative use cases from industrial practice.

It does not cover liquid coatings, inks, or solvent-based systems, only melt-processing of solid thermoplastics.

2. Key Functions

Dispersing and wetting additives deliver measurable performance benefits across compounding and conversion processes. Their primary functions are not just theoretical, they directly impact yield, quality, and cost. The key practical roles include:

2.1 Enhancing dispersion of pigments, fillers, and reinforcements

Dispersing pigments, fillers and other reinforcing agents is a fundamental need in thermoplastic compounding, not only to provide visual appeal, but also to provide consistency in mechanical, thermal, and processing performance. Agglomerates are formed by poor dispersions of additives because the interparticle force (e.g., van der Waals attraction) is a stress concentrator and contributes to the interfacial adhesion and degrades important properties like impact strength tensile modulus and fatigue resistance. Incomplete dispersion causes irregular hue, diminished color strength, and surface flaws such as specks or streaks; a problem particularly sensitive in automotive, packaging, and consumer goods, where aesthetics and practicability are highly important. The direct solutions to these issues are dispersing and wetting additives, which enhance the separation of agglomerates during the melt mixing process as well as the uniform distribution of individual particles in the polymer matrix.

Effective dispersion also improves process efficiency and flexibility in formulation besides aesthetics and mechanical integrity. Dispersed systems have lower viscosity of the melt and have lower torque in extrusion, allowing greater throughput and lower operational energy, and they are also smoother to operate, especially with highly filled compounds (e.g. $CaCO_3$ in HDPE pipe or talc in automotive parts made of PP). When using reinforced material like glass-or carbon-fiber composites, the proper wetting of fibers reduces the breakage of the fibers and enhances the transfer of load between the reinforcement and the matrix which also has a direct effect on the performance of the part. In the final analysis, the purpose of dispersing and wetting additives is not only to mix better, but to release the full functional utility of all solid additives in the formulation so that conceptual design of compounds becomes predictable and consistent performance results.

2.2 Enhancing masterbatches with color strength, consistency and gloss

High color strength, batch-to-batch uniformity, and surface gloss cannot be accomplished by pigment choice alone in color masterbatch production, but rather determined by the capacity of pigments to be wetted and dispersed in carrier resin. Inadequate dispersion is such that pigment agglomerates remain, and these do not disperse light evenly, they weaken tinting and cause visual imperfections like mottling or specking. By reducing the interfacial tension between the surface of the pigment and the melt polymer, dispersing and wetting additives have a decisive role in allowing full wetting and effective deagglomeration of the compounding process. This maximizes the amount of exposed surface of the pigment particles, which directly increases color development and enables formulators to develop target shades with lower pigment loadings-better cost and process stability.

In addition to the raw color strength, these additives are needed to guarantee reproducibility and surface quality in the downstream uses. Constant dispersion reduces changes in Delta-E between production cycles, which is important in production processes such as the automotive, packaging, and consumer electronics where brand color standards are stringently regulated. Also, a dispersed system helps in better flowing of the melt and minimizing roughness in the surface of the final part, which transfers to glossiness and improved aesthetics, particularly in thin-wall injection molding use or cast film. The stabilizing effect of dispersants in high-shear processes such as twin-screw extrusion also eliminates re-agglomeration of pigments during residence time, and entraps uniformity between pellet and part. This does not only produce a product that is visually appealing, but it also produces a product that is of high quality and has reduced number of rejects and rework.

2.3 Reducing agglomeration, screen pack clogging, and downtime

Agglomeration of pigments, fillers or other solid additives is not just a nuisance to formulation in thermoplastic compounding and extrusion, but it is a direct translation of operational inefficiency and higher production costs. Particles that do not soak well or as individual micro-aggregates serve as physical barriers at the melt filtration process. This causes quick accumulation on the screen packs and this results in a continuous increase in the pressure drop across the filtration zone. Uncontrolled, this causes unscheduled interruptions to line to change screens, and interferes with the homogeneity of the melting process and can even cause damage to gear pumps or extruder parts because of too much backpressure. This is counteracted by dispersing additives and wetting additives that allow the primary particles to be completely separated and encapsulated at the earliest stage of the melt stage to limit the appearance of hard agglomerates that are difficult to break down by shear and contribute to filter fouling.

Decreased agglomeration does not only have an influence on a smoother extrusion but improves reliability of the processes in general as well as the utilization of assets. When compared to stable dispersions, stable dispersions ensure the same pressure profile during extended run times allowing extended production run times and minimizing labor and materials wastage due to purging and startup/shut down procedures. Even a 1015% increase in screen life can be a cost savings in high-output processes like pipe extrusion, film blowing, or masterbatch manufacturing, and can increase a company's performance on-time delivery. Also, reduced interruptions will ensure improved product quality since instantaneous changes in pressure and flow instabilities (when changing the screen) are eliminated. Processors reduce a frequent maintenance bottleneck into a manageable, predictable variable by handling dispersion at the design phase, which directly serves the lean manufacturing and operational excellence objectives.

2.4 Boosting surface appearance and mechanical integrity

It is not a mere cosmetic objective to attain high surface appearance in plastic components, but more frequently it is a clear measure of the dispersion quality and interfacial compatibility. Inadequately dispersed additives, be it pigments, fillers, or reinforcements, form microscale heterogeneities that are exhibited by visual defects such as specks, streaks, orange peel or dull spots especially in high-gloss or thin-wall applications. The solution to these is the dispersion and wetting additives which can guarantee the complete wetting and even distribution of solid particles during the melt processing. This causes the melt flow to be smoother, the surface roughness to be less pronounced and the light reflection to be uniform- important to automotive interiors, consumer electronics housings and packaging where appearance is the initial factor to attract customer acceptance. Even sub-visible agglomerates may cause scattering in transparent or semi-transparent systems and compromise clarity; in wetting that effectively minimizes this aspect, optical performance is not compromised to achieve functionality.

In addition to the look, the mechanical integrity of a formed part is directly linked to the degree of additives incorporation in the polymer structure. This is because agglomerates are stress concentrators, which cause microcracks during the load application, greatly compromising its impact strength, elongation at break, and fatigue capacity. In filled structures, e.g. with talc-reinforced PP or glass-fiber PA, wetting of the interfaces are poor, which results in a weak bonding between the filler and the resin, which is the negation of the entire point of reinforcement. Dispersing agents increase adhesion promoting close contact of the phases allowing stress to be transferred efficiently and to get the maximum contribution of each additive to the performance of the composite. The outcome is a part that is both visually cleaner and has the reliability to produce parts that meet or even surpass mechanical design specifications across production cycles, low scrap, warranty and/or field failures.

2.5 Lowering melt viscosity and improving processability

Reduction of melt viscosity is a significant but forgotten advantage of optimally selected dispersing and wetting additives in thermoplastic processing. When including a solid filler, particularly in very-filled systems like $CaCO_3$-filled polyolefins or glass-fiber-reinforced engineering resins, solid additives raise internal friction, flow resistance and induce higher torque, higher motor load, and possible thermal damage due to excessive shear heating. The mitigation of this is through dispersing agents that enhance efficient particle wetting as well as minimize interparticle friction, which in effect, work as internal lubricants that drive the polymer chain around dispersed phases through smoother pathways. This causes a quantifiable reduction in melt viscosity (which is often manifested in higher Melt Flow Index), allowing processors to operate at lower temperatures, use less energy, or higher screw speed with no loss in material stability.

On the production perspective, enhanced processability is directly proportional in terms of operational flexibility and control of costs. Reduced viscosity improves the loading of fillers without compromising extrusion stability, enables a thinner-walled or complicated geometry to be injection molded, and minimizes the risk of melt fracture in film or fiber processes. Viscosity reduction is required in masterbatch production where the concentrations of pigments may be very high of up to 40 and 60; otherwise, the extruder would be overloaded and homogenizing would be necessary. Additionally, the uniform rheology between batches reduces variability in start-up and promotes automation friendly processing windows. Finally, the viscosity-modifying mass of the dispersing and wetting additives is not simply of the easier flow, but rather is of enabling the manufacture of formulations that would otherwise be in practice only at small scale, without affecting throughput, quality, and equipment life.

2.6 Supporting compatibility in multi-phase systems

Throughout multi-phase polymer systems (i.e. blends of polyolefins and engineering resins (e.g. PP/PA), polymer/filler composites, or wood-plastic composites (WPCs)) attaining interfacial compatibility remains a thorny technical issue. The nature of these systems is that there is a bad adhesion between the dissimilar phases because of the incompatibility of the surface energies and results in the separation of the phases, weak boundaries, and reduce the mechanical performance. The addition and wetting of additives is a very important step since it serves as an interfacial modulator: interfacial tension between the polymeric matrix and the secondary phase (between the polymer and another polymer, mineral filler, natural fiber, etc.) is lowered, enhancing wetting and closer contact. This has an increased compatibility that reduces the formation of voids, enhances the transfer of stress across interfaces and stabilizes the morphology in high shear melt processing.

Processing wise, this interfacial stabilization will be translated into practical advantages: delamination in the co-extruded layers will be reduced, surface flaws in filled compounds will be minimized and fiber dispersion in WPCs or glass-filled nylons will be more even. With reactive or non-reactive polymer blends, some functionalized dispersants (e.g. maleated polyolefins such as PE-g-MA) may even exhibit some level of coupling effect, which chemically bonds the phases, further improving toughness and thermal stability. In the end, it is not only that use of custom wetting agents in multi-phase formulations prevents separation, but allows creation of hybrid materials with the desired properties of each constituent (e.g., stiffness due to fillers, ductility due to elastic, or barrier properties due to polar) without compromising processability or final-use performance.

3. Mechanisms of Action

The performance of dispersing and wetting additives in thermoplastic processing stems from three interrelated physical mechanisms: wetting, dispersion, and stabilization. While often discussed separately, these steps occur sequentially, and sometimes simultaneously, during melt compounding. Understanding how each mechanism functions helps in selecting the right additive for a given polymer-additive system.

3.1 Wetting Mechanism

The basic procedure in the dispersion of solid additives, which could be pigments, fillers, or reinforcements, into an effective dispersion in a thermoplastic matrix is the wetting mechanism. These particles, in their dry form, may generally be covered by air or adsorbed water and have high surface energy, particularly with the inorganic substances of calcium carbonate, talc, or titanium dioxide. Non-polar polymers like polyethylene (PE) or polypropylene (PP) on the other hand have low surface energy which forms a thermodynamic mismatch that prevents spontaneous wetting. The molten polymer will not be able to penetrate the particle surfaces and cover them completely without the intervention, creating voids or dry spots where the agglomeration can begin and defects occur. This is overcome by dispersing and wetting additives which adsorb onto the particle surface and lower the interfacial tension between the solid additive and the polymer melt which priming the interface so that it is now subject to intimate contact.

Processing-wise, effective wetting would be translated into quicker incorporation of the wet end when melt compounding is done and lowers the amount of mechanical energy needed to obtain homogeneity. Strongly-affinity additives to the surface of the particle (polar and functional groups) and the polymer matrix (compatible hydrocarbon chains) serve as molecular bridges and allow the polymer melt to replace air, and completely encasements each particle. This does not only discourage the hard agglomerates but preconditions the further deagglomeration and stabilization. In practice, good wetting reduces such problems as specking, incomplete color development, or uneven distribution of fillers, especially important in high-speed extrusion or in thin-wall molding when the residence time is brief and the process tolerances are narrow. Finally, the wetting mechanism is not only a chemical phenomenon, it is a determining aspect in whether a formulation works out or remains the cause of persistent headaches during production.

3.2 Dispersing Mechanism

The next phase of action is dispersing, which goes hand in hand with the deagglomeration of the solid additives such as pigments, mineral fillers, or reinforcing fibers in melt compounding. The mechanical shear provided by the extruder screw is responsible for most of the power needed for deagglomeration; however, it is often not enough for very fine and high-surface-area solid additives with strong cohesive forces. Dispersing and wetting additives break through the primary bounds of agglomerates and generally help the internal interactions that hold agglomerates together to die. In most cases, the molecular structure of their polar or functional head group is capable of binding to the particle's rivet surface, while the non-polar tail can be dissolved in the polymer matrix, thus granting them the ability to function as "molecular wedges," aiding the separation of particles under practical processing shear levels.

In practical scenarios, the degree of dispersion directly affects whether the formulation will function smoothly or be plagued by persistent defects. Even a high-intensity twin-screw extrusion may fail to entirely break micro-agglomerates if dispersing action is not sufficient, thus leading to these micro-agglomerates later showing up as specks in film, streaks in injection-molded parts, or weak points in structural components.

The success of this mechanism is conditioned by several factors: the migration rate of the additive to the particle interface, its thermal stability at processing temperature, and its compatibility with both the solid phase and the base resin. As an example, low-molecular-weight waxes (e.g., PE wax) quickly wet pigment surfaces in polyolefins, whereas functionalized polymers like maleated polyethylene (PE-g-MA) provide more robust anchoring in highly filled mineral systems. In the end, the dispersing mechanism is what helps to go beyond laboratory formulation and achieve production that is stable and reliable, thus making it possible for every particle to contribute to the final product and not be a source of defects.

3.3 Stabilization Mechanism

After pigments, fillers, or reinforcements are wetted and deagglomerated, the dispersion has to be kept throughout the rest of the melt process and in the final product; otherwise, particles will re-agglomerate because of collisions during flow or cooling. At this point, the stabilization mechanism is very important. Different from liquid systems (for example, paints or inks), where electrostatic repulsion can prevent particle re-association, thermoplastic melts are non-conductive, highly viscous, and do not have mobile ions, hence electrostatic stabilization is not effective. Polymer processing uses steric stabilization as the main and practically feasible mechanism. The molecular architectures of the additives that are designed for this role have an anchoring segment that strongly binds to the particle surface and a compatible "tail" (for example, polyolefin chains in PP or PE) that goes on the surrounding melt. This tail forms a physical barrier that keeps the closest particles from coming into contact with each other again for van der Waals forces to pull them together.

Good steric stabilization in industrial compounding means direct and consistent product quality throughout different production runs. Even a thoroughly dispersed melt might result in quality degradation during the downstream operations if steric stabilization is not present, these could be strand pelletizing, injection molding, or film extrusion where shear history and thermal cycles differ. At this point, re-agglomeration results in specks that appear later, color intensity that is not uniform, or local stress points that lower mechanical performance. So, the stabilizing agent should not only match the base resin's chemistry but also the processing window: a dispersant with polyethylene tails will stabilize $CaCO_3$ in HDPE well, but may separate phases in polycarbonate. Besides that, thermal stability is equally important, the additive should not evaporate or degrade, but rather stay firmly anchored and functional even at the peak processing temperature. In the end, stabilization is not only about saving the dispersion; it is also about being able to recall the performance that was achieved during wetting and deagglomeration, hence, the performance is assured from the extruder to the end-use part.

4. Types & Examples

Dispersing and wetting agents cover an extensive variety of chemical families, each one designed for different polymer systems, additive types, and processing conditions. Choosing the appropriate one involves not only knowing the chemical structure but also the functioning of the agent under melt conditions. Seven main categories that are commonly used consist of polymeric waxes, fatty acid derivatives, metal soaps, surfactant-based systems, functionalized polymers, silicone additives, and amide-based dispersants. The following is a functional overview of each type along with typical formulations, performance features, and application fields.

4.1 Waxes

Polymeric and natural waxes are the most popular dispersants in polyolefin systems due to their great compatibility, cheapness, and their dual role as both dispersants and internal lubricants. Polyethylene wax (PE wax) and polypropylene wax (PP wax) are very effective in PE and PP matrices, respectively, because their molecular structure is very similar to that of the carrier resin, thus enabling them to adsorb strongly on the pigment and filler surfaces. Montan wax (a natural hydrocarbon wax obtained from lignite) provides higher thermal stability and is, therefore, more frequently used in high-temperature compounding or rigid PVC areas. These waxes lower interfacial tension, enhance particle wetting, and reduce melt viscosity thus making them very important in masterbatch and mineral-filled compound production.

4.2 Fatty Acid Derivatives

Fatty acid derivatives such as stearic acid and ethylene bis-stearamide (EBS) are very flexible, cheap dispersants that can be used in a broad range of polar and non-polar systems. Industry-wise EBS is the most important additive–it works as a mild dispersant, external lubricant, and anti-blocking agent. Its amide structure provides the ability of being migrated at interfaces during processing and thus it reduces the friction between particles as well as between the melt and metal surfaces. EBS is a general-purpose compound, PVC, and low-cost masterbatch production where moderate dispersion performance is needed and, in such cases, it is a staple though less than very specialized dispersants is it.

4.3 Metal Soaps

Metal soaps, like calcium stearate, zinc stearate, and magnesium stearate, are mainly employed as processing aids and stabilizers in PVC and polyolefins but also help pigment wetting and flow by their slight interfacial activity. Because of their ionic character, the metal soaps become moderately interfacially active especially in polar conditions. For example, calcium stearate is a common rigid PVC additive that helps the dispersion of heat stabilizers and pigments apart from it also acts as a lubricant. Zinc stearate, though very efficient, must be implemented carefully in polyolefins to avoid the metal acting catalytically on oxidative degradation at high temperatures. These are the best solutions for moderate-dispersion-demand scenarios and are often found in mixtures with stronger dispersants in heavy-duty applications.

4.4 Surfactant-Based

Surfactant-based components like sorbitan esters (Span®), polyethylene glycol (PEG) esters, and ethoxylated fatty alcohols are very efficient in polar polymer systems like PET, PA, PC, and PVC where non-polar waxes have poor compatibility with them. These agents' amphiphilic nature - having a hydrophilic head and lipophilic tail - permits strong anchoring to high-energy surfaces such as TiO_2 or carbon black and at the same time that they are soluble in polar melts. These agents are excellent in color masterbatches which require high gloss and color strength, mainly in engineering plastics. Nevertheless, they are, in general, not suitable for polyolefins due to poor compatibility as well as possible migration and blooming. Their thermal stability is also an issue since most of them decompose above 220°C thus being limited in high-temperature processes.

4.5 Functionalized Polymers

Functionalized polymers are the most high-performing class of dispersants for most demanding application scenarios. Among them are maleic anhydride grafted polyolefins (e.g. PE-g-MA, PP-g-MA), ethylene-vinyl-acetate (EVA), and other polar-modified resins ones. Their main purpose is to be chemically linked both with the polymer matrix and the additive surface - thus acting as real compatibilizers - which is the underlying principle of their immense value. For example, in mineral-filled systems (e.g., CaCO3 in HDPE), PE-g-MA helps the connection between the two phases, thus lowering the agglomeration and improving mechanical characteristics. In fiber-reinforced composites (e.g., glass in PBT) they reduce the breaking of fibers while enhancing the transfer of stress. Though they are more expensive than waxes or EBS, functionalized polymers are capable to deliver a measurable performance uplift and this is why they are indispensable for the high-end automotive, electrical, and structural sectors.

4.6 Silicone-Based

The main reasons for which silicone-based dispersants, mostly brought in as PDMS (polydimethylsiloxane) masterbatches, are accommodated in the industry are their remarkable surface activity and flow enhancement capability. They are capable of very drastically reducing melt viscosity as well as surface tension and in this way, carbons black or nano-fillers are very thoroughly wetted although are difficult to disperse. Because of their low surface energy, they also make the demolding and surface-gloss better and for this, they are favorites in high-clarity applications and thin-wall injection molding. Nevertheless, silicones may cause problems with printing, painting, or bonding if they relocate to the surface of the part. Their use implies that dosage as well as processing conditions should be controlled carefully so as to prevent plate-out or contamination. These compounds are mostly potent in polyolefins as well as engineering thermoplastics where the quality of the surface is very important.

4.7 Amide-Based

Amide-based components, for instance, erucamide and oleamide, are long-chain fatty amides that mainly work as external lubricants and slip agents, though they also help pigment wetting in polyolefins. Their straight structure enables them to move to the polymer surface during processing; thus, friction is lowered and the flow is improved. Even though they are less powerful than waxes or functionalized polymers when it comes to breaking up agglomerates of fine pigments, they are commonly utilized as a mixture with other dispersants to improve the overall processability. In LDPE and HDPE films and containers, which are the most typical applications, they assist in the prevention of die build-up and in the increase of extrusion rates. Owing to their low melting point (~100–120°C), their use is limited to high-temperature systems, however, they are still quite useful in the low to mid-range of compounding where surface finish and processing speed are the main factors.

Table 4.7: Types of Dispersing & Wetting Additives: Examples and Typical Uses

Type	Examples / Chemical Basis	Typical Use
Waxes	PE wax, PP wax, Montan wax	Pigment/filler dispersion in PE, PP
Fatty Acid Derivatives	Stearic acid, EBS	Lubrication and dispersion
Metal Soaps	Calcium stearate, Zinc stearate	Pigment wetting
Surfactant-Based	Sorbitan esters, PEG esters	Pigment wetting in polar polymers
Functionalized Polymers	PE-g-MA, EVA	Filler coupling and dispersion
Silicone-Based	PDMS masterbatches	Wetting and flow improvement
Amide-Based	Erucamide, Oleamide	Improved pigment wetting

5. Selection Criteria

Dispersing and wetting additives selection is not a scientific experiment with random trials. It is a scientifically driven decision that depends on aligning the additive's chemical and physical properties with the polymer matrix, filler/pigment system, and processing conditions. If the components do not fit, the result may be ineffective dispersion, plate-out, degradation, or even product failure. Below are the five essential criteria that should be quantitatively assessed before the formulation is finalized.

5.1 Selection Criteria by Polymer Polarity

Additive selection is most heavily influenced by the polarity of the base resin. Non-polar polymers, like polyethylene (PE), polypropylene (PP), and polystyrene (PS), have a low surface energy and do not have functional groups, thus they are not compatible with highly polar or ionic dispersants. Non-polar or weakly polar additives such as polyethylene wax, EBS, or silicone masterbatches give the best results in these systems because their hydrocarbon chains are almost identical to the matrix, thus ensuring good miscibility and minimal migration.

Meanwhile, polar resins (Polyamide (PA), polycarbonate (PC), polybutylene terephthalate (PBT), and polyethylene terephthalate (PET)) have the ester, amide, or hydroxyl groups that can strongly interact with polar or functionalized dispersants. In this case, surfactant-based systems (e.g. PEG esters, sorbitan derivatives) or functionalized polymers like PE-g-MA or EVA are much effective that they interact by dipole interactions or hydrogen bonding with both polymer and additive surface. A non-polar wax in PA or PET usually brings blooming, poor dispersion, and mechanical properties degradation. Always ensure that the polarity of the additive matches that of the resin: a mismatch will weaken every other performance metric.

5.2 Thermal Stability Considerations

Dispersing agents should be able to withstand the processing environment's thermal history without decomposing, volatilizing, or altering chemically. Most polyolefin compounding is done between 180°C and 230°C; the temperature for engineering plastics like PBT or PC may be 260–300°C. Many low-molecular-weight waxes and surfactants start to decompose or evaporate when the temperature is above 220°C, which results in the loss of the agent's efficacy, odor, or smoke, especially in the closed-loop extrusion systems. In the case of high-temperature, choose additives that are thermally stable: Montan wax (decomposition >250°C), functionalized polymers (e.g., PE-g-MA, stable up to 280°C), or silicone-based masterbatches that are intended for use with engineering resins. To be totally certain that the additive is still chemically intact throughout processing, one should not consider only the melting point but also the supplier's TGA (thermogravimetric analysis) data.

5.3 Temperature Sensitivity

The performance of the additive may get affected dramatically by the temperature gradients even if it is within a stable processing window. For instance, in twin-screw extruders, while the metering zone may be at 240°C the feed zone is probably at 160°C. Some additives, mainly amide-based lubricants such as oleamide or erucamide, may dissolve or soften at low temperatures (100–120°C), which results in the early migration to the screw or barrel wall before full dispersion takes place. This leads to inconsistency in wetting and poor coverage of particles. On the contrary, high-melting waxes such as Montan wax may only become active at higher temperatures, thus delaying the beginning of dispersion. The right additive is the one which, at the moment the polymer melts, gets activated, usually within 10–20°C of the resin's melting point, so that wetting can be done on time and migration can be avoided. For exact control, additives with narrow melting ranges or those delivered as masterbatches for controlled release are to be used.

5.4 Shear Rate Effects on dispersion efficiency

Dispersion is a very indirect process, it is a dynamic communication, between the chemistry of the additive and the mechanical energy input brought about by the process. Low-shear systems (e.g., batch mixers, single-screw extruders) implement additive functionality to a great extent in force reduction between particles as mechanics partly deagglomerates the system. Thus, in these cases, it is absolutely necessary to use the most potent on the surface agents such as highly functionalized polymers or high-polarity surfactants that achieve chemical interaction at the surface.

On the other hand, high-shear twin-screw extruders are able to produce a strong mechanical energy that perform physical breaking of the agglomerates thus the additive play is more about stabilization than deagglomeration. Therefore, here, low-viscosity waxes or silicone additives which are capable of reducing the melt's friction and thus facilitating its further breakup as well as preventing the re-agglomeration of particles after the shear step are more useful than high-molecular-weight stabilizers. By using too much shear without proper wetting, one can end up with "false homogeneity" - the particles may appear to be evenly dispersed to the eye, but in reality, they are internally clustered, which eventually results in late-stage defects. The main point is ensuring a proper match between the function of the additive and the shear regime: chemical wetting should be used in low-shear, and steric stabilization in high-shear.

5.5 Residence Time Requirements

Residence time is the length of time the material is in a molten state and it is a decisive factor in whether dispersion is thorough and stable. In high-output extrusion lines (e.g., pipe or film), residence time can be as short as 1–2 minutes. On the other hand, in batch compounding or slow injection molding cycles, it can be more than 10 minutes.

Extremely short residence times require the use of fast-acting additives: low-molecular-weight waxes, EBS, or silicone-based systems that are able to quickly migrate to particle surfaces and stabilize them immediately. The slower performance of functionalized polymer or high-MW dispersants may result in incomplete dispersion due to the short time allowed for adsorbing and anchoring. On the contrary, in situations where the residence time is long (e.g., compounding for masterbatches that will be used in the downstream extrusion process), the preference should be given to the slower but more enduring stabilizers, such as PE-g-MA, because they provide the resistance to re-agglomeration that lasts both during storage and reprocessing. It is always important not only to think about how the mixing is done but also how long the material will stay molten. A dispersant that works perfectly in a laboratory extruder may not be suitable for a production line with half the residence time.

The decision framework utilized by top compounders to remove the element of guessing is outlined below. In choosing a dispersant, consider the following:

1. Is it compatible with the polarity of my resin?
2. Is it going to be able to withstand the temperature of my process?
3. Is it going to be activated at the correct point in the melt profile?
4. Is its function consistent with my shear environment?
5. Am I having enough time for it to work?

Systematically answering these five questions is a guarantee that your additive selection will be a source of real performance rather than a mere promise.

6. Typical Applications

Dispersing and wetting additives are not just simple additives; they are sophisticated tools designed to overcome specific challenges in different processing environments. Their impact can be significantly different depending on where they are used, the polymer system, and the set of performance criteria. The following is a description of the main application areas in which these additives bring the measurable value of their performance.

6.1 Color Masterbatches

In the production of color masterbatch, which is the case of high-demand sectors like automotive, packaging, and consumer electronics, dispersion quality is what mainly determines customer acceptance. Slight specking or color variation may result in the rejection of the whole batch. Dispersing agents here facilitate the production of the most complete wetting of ultra-fine pigment particles (e.g., carbon black, phthalocyanines, or organic reds), thus, tinting strength is maximized and dosage is minimized. Not only does this contribute to lowering the cost, but also enhancing gloss and surface smoothness of injection-molded or extruded parts. In high-loading systems (e.g., >15% pigment), steric stabilizers such as functionalized polymers or silicone masterbatches help to prevent re-agglomeration during pelletizing and storage, thus, color consistency from batch to batch is assured.

6.2 Filler Masterbatches ($CaCO_3$, Talc, etc.)

Mineral-filled compounds are very common in packaging, automotive, and construction industry; however, the high filler loadings (most of the time 30–60%) greatly increase melt viscosity and create severe dispersion challenges. Without the proper wetting, $CaCO_3$ or talc particles cluster together forming hard agglomerates which act as stress concentrators; thus, impact strength is lowered, and die build-up or screen clogging occur. Dispersants like polyethylene wax or PE-g-MA separate the particles better, thus, the torque during the extrusion is lowered, and the adhesion between filler and the polymer is improved, resulting in easier processing, higher speed of the process, and better mechanical properties. For HDPE pipes or PP automotive parts, dispersion at the optimum level can bring about the reduction of shrinkage and warpage due to the uniform thermal expansion.

6.3 Engineering Plastics Compounding (PA, PC, PBT)

Usually, the reinforcements of engineering resins are glass fibers, carbon fibers, or flame retardants, fibers that are abrasive, high-surface-area, and very hard to wet. Poor dispersion brings about the breakage of fibers, inconsistent flow, and weak interfacial bonding; thus, mechanical strength is compromised. Dispersing agents in such systems are the reduction of melt viscosity agents that allow better flow to thin-wall molds, and on the same time, they protect the fibers' surface from damage during compounding. Functionalized polymers such as PA-g-MA or EVA are extremely important in this case, they couple chemically with polar polymer chains while simultaneously wetting hydrophilic fibers, thus, stronger, more durable composites are obtained. In electrical connectors made of PBT, the fiber clumping even at a very low level can cause electrical tracking or warpage, thus, dispersion is not just a processing issue, but also a reliability one.

6.4 PVC Processing

PVC recipes are the mixture of stabilizers, lubricants, impact modifiers, and pigments, all that require high shear, intimate mixing, and moderate temperature. Dispersing agents facilitate the suspension of heat stabilizers (e.g., organotin or Ca/Zn systems) and stop their agglomeration, which otherwise can result in localized degradation and black specks. In rigid PVC, metal soaps like calcium stearate are both dispersants and lubricants while in flexible PVC, surfactant-based agents facilitate the dispersion of plasticizers and pigments. The problem here is the dispersion-lubrication balance, too much internal lubrication can lead to melt fracture; too little can result in die lines or poor surface finish.

6.5 Rubber & TPE Compounds

Dispersing agents in TPEs and rubber compounding are the most important ones when incorporating fillers like carbon black, silica, or clay, as well as softening agents and curatives. Inefficient dispersion thus leads to inconsistent cure rates, poor tensile properties, and stickiness of the surface. In TPEs based on SEBS or PP/EPDM blends, waxes and amide-based additives assist the flow in extrusion and injection molding and at the same time, they provide the even distribution of oil-based softeners. Thus, blooming is prevented and dimensional stability is kept, the two parameters that are most important for automotive seals, grips, and consumer product housings.

6.6 Wood-Plastic Composites

WPCs are the material which combines lignocellulosic fibers (wood flour, bamboo) with polyolefins in order to produce the decking and fencing materials that are durable and of low maintenance. What is the problem? The wood fibers are hydrophilic and have a poor affinity for the hydrophobic plastics, and that is the reason behind poor adhesion, void formation, and moisture absorption. Dispersing agents, especially maleated polyolefins (PE-g-MA, PP-g-MA), are the coupling agents that form ester linkages with hydroxyl groups on the wood surface while simultaneously bonding with the polymer matrix. The process not only changes the dispersion but it also increases the moisture resistance, decreases the water uptake, and raises the bending strength. The WPCs without these additives experience delamination, roughness of the surface, and premature failure in the outdoor environment.

Each of these applications requires a different solution, not only in the choice of the additive but also in formulation strategy. A dispersant that performs excellently in a color masterbatch may not work at all, or even negatively, in a WPC or PBT compound. The main thing is not to know what the additive does, but to know how it interacts with the whole system: polymer, filler, processing equipment, and end-use environment.

Table 6.6: Application Areas and Role of Dispersing/Wetting Agents

Application Area	Role of Dispersing/Wetting Agent
Color Masterbatches	Improves pigment dispersion, gloss, and color uniformity
Filler Masterbatches ($CaCO_3$, Talc, etc.)	Enhances filler distribution and reduces melt viscosity
Engineering Plastics Compounding (PA, PC, PBT)	Reduces fiber breakage and enhances flow
PVC Processing	Aids dispersion of stabilizers and pigments
Rubber & TPE Compounds	Promotes uniform distribution of fillers and softeners
Wood-Plastic Composites	Improves adhesion and dispersion of lignocellulosic fibers

7. Application Methods

How a dispersing or wetting additive works cannot be figured out just by looking at its chemical structure - it is also very much dependent on how the additive is brought into the polymer matrix. The application way significantly affects the dispersion efficiency, the stability of the process, and ultimately the product quality. There is no one universal 'best' way of doing it; the best approach depends on the polymer system, the kind of additive, the scale of production, and the configuration of the equipment. The following are the most common and industrially proven application methods, accompanied by their advantages, restrictions, and guidelines for implementation.

7.1 Dry Blending (Pre-Mixing)

Dry blending is the method of physically mixing a solid additive (for instance, a pigment, a filler, or a dispersant) with the polymer resin in a powder or pellet form and doing this before the mixture is fed into the extruder or the mixer. This is the main method for the production of masterbatch and compounding of low-to-medium volume.

Advantages:

- It is simple, low-cost, and most extruders and batch mixers can use it without any problems.
- It allows additive ratios to be controlled very precisely and custom blends can be pre-formulated.
- It is a method that does not expose sensitive additives to thermal stress excessively (e.g., surfactants or amides) before melting.

Best Practices:

- Do not settle for ordinary mixers; use high-shear tumble blenders or ribbon mixers in order to achieve uniformity - this is particularly important when you are mixing finely ground pigments with coarsely grained pellets.

- Always blend the dispersant with the pigment/filler first and then add the polymer. Doing so, the dispersant will be the one that covers the particles of the solid before they come into contact with the polymer melt.
- Prevent static from building up when dealing with conductive additives (e.g., carbon black) by using anti-static agents or grounded equipment.

Limitations:

- If the mixing is not complete it can cause local areas where overdosing or underdosing happen, thus specking or inconsistent color may be generated.
- The method is not suitable for ultra-fine nanoparticles or for highly cohesive fillers unless additional energy is provided.

7.2 Masterbatch Incorporation

Masterbatches are the concentrated mixtures of additives (usually 20–60% of the active ingredient) homogenized in a carrier resin (e.g., LDPE, PP, or EVA). After that, the masterbatch is diluted in the final polymer during extrusion or injection molding.

Advantages:

- It very well guarantees the dosing to be consistent and repeatable, also at very low additive levels (e.g., 0.1–0.5%).
- By lowering the dust, the methods improve the safety of the handlers and the risk of contamination is minimized.
- It makes feasible the implementation of additives that are not compatible with the final polymer (e.g., surfactant-based dispersions in PP with a PE carrier).

Best Practices:

- For the best compatibility, the masterbatch carrier resin should be the same as the final polymer (e.g., use PP carrier for PP parts).
- Keep a steady dilution ratio (e.g., 5:1 or 10:1) and make sure the feeding rates are correct so as not to have metering errors.
- For highly demanding uses (e.g., automotive color), double-screw extruders should be used to ensure the masterbatch is fully dispersed during dilution.

Limitations:

- It takes more money because of the additional step of the process.
- If masterbatches are poorly prepared, they can re-agglomerate the areas where the initial dispersion is insufficient.

7.3 Direct Addition During Extrusion

The method here is that the dispersing agent is the one that is brought directly into the feed throat or metering zone of the extruder along with the base resin and solid additives. The process is characterized by a high-volume, continuous setup like film blowing or pipe extrusion.

Advantages:

- Pre-blending steps are removed which is a saving of labor and inventory.
- It is very suitable for dynamic formulations where additive levels vary from one product to another.

Best Practices:

- Liquids or pelletized dispersants should be taken to the melting zone (after the feed throat) rather than the feed zone if you are willing to stop the premature migration or volatilization.
- For accurate dosing, gravimetric feeders should be employed, dosing is very critical for systems with low additive levels (e.g., 0.2–0.8%).
- In the case of powder additives, air-assisted feeders should be used so as to avoid bridging and to ensure that the flow is constant.

Limitations:

- There need to be strong mixing elements (e.g., kneading blocks) after the point of injection in order to get the mixing done properly.
- High shear zones are mandatory; this method cannot be used in a low-shear single-screw system which does not have a downstream compounding unit.

7.4 Liquid Additive Feeding (for Oil-Based or Solution Systems)

Several dispersants, especially the silicone-based or the surfactant systems, can be delivered only as liquids or in solution form. In most cases, these are metered by pump systems straight into the melt stream.

Advantages:

- Allows almost ultra-fine dosing (up to 0.01% levels).
- This is the best option for lines with very high speed where it is not feasible to feed solids.

Best Practices:

- Lines and nozzles that are heated should be used so as to keep the viscosity and prevent clogging.
- It is also very important to always do the injection in a high-shear zone (e.g., after the melting zone) so the dispersion can be very quick.
- Always check for compatibility, some liquid additives if injected too early or into incompatible resins can give rise to phase separation.

Limitations:

- It is also equipped with very specialized metering instruments and needs constant maintenance.

- There is always a chance of spillage, evaporation, or even fire when dealing with volatile solvents (although most modern systems use solvent-free liquids).

7.5 In-Line Dry Powder Injection (for High-Load Filler Systems)

For instance, in mineral-filled HDPE or PP (e.g., 50% $CaCO_3$), the common practice is to inject the dry filler powder straight into the extruder barrel through the side feeders. The dispersants are usually added at the same time through a different feed port.

Best Practices:

- Synchronized feeding should be practiced: just before or at the same time with the filler, the dispersant should be introduced so that it can immediately wet the filler.
- Keep a strong vacuum at the side feeder so that no air can be sucked in and no moisture can be absorbed.
- Immediately downstream, combine with high-shear mixing elements (kneading discs, reverse elements).

Critical Insight:

- "The dispersant in a high-filler system is not only mixing helper, it is process enabler. Without it, the melt is getting too thick to be pushed through the pipe, and the filler is just clogging the screw."
- Each method is used in a different place and situation in industrial practice. The top compounders don't put their faith in one method only; they are tactical in combining them. For instance:
- Dry blend dispersant + pigment → Masterbatch → Direct feed into extruder with filler.

The objective is not to merely add the additive, but also to make sure that the particle surface is getting the additive at the right time, in the correct state, and with enough energy for it to work. Using the wrong method can make your dispersant, even if it is the best one, totally ineffective. Always validate your method with a filter pressure rise test or speck count before full-scale production.

8. Processing Conditions

Dispersing and wetting additives are not only a matter of chemistry; they depend significantly on the method of processing the melt. The wrong temperature profiles, insufficient shear, or inadequate residence time can cause even the most advanced dispersant to fail. Optimal processing conditions differ greatly for different polymer systems, additive types, and applications. The following is a practical guide to setting up extrusion and compounding parameters for each major application, based on actual plant data and industry best practices.

8.1 Color Masterbatches

High pigment loadings demand more than just mixing, they require true dispersion. In color masterbatch production, especially for automotive, packaging, and consumer electronics, even slight agglomeration leads to specks, inconsistent shade, or poor gloss. This section provides field-validated processing profiles that balance temperature, shear, and residence time to ensure full wetting of carbon black, TiO_2, and organic pigments, without degrading sensitive additives like EBS or surfactants. It's not about running hotter or faster; it's about optimizing every zone so the additive does its job before the polymer leaves the extruder.

Target: Get complete pigment wetting without thermal degradation or specking.

Temperature Profile:

- Feed Zone: 140–160°C (to keep the pellet from melting prematurely and thus avoiding bridging)
- Melting Zone: 180–200°C (PE/PP); 210–230°C (PVC, PA)
- Metering Zone: 200–220°C (PE/PP); 230–250°C (engineering resins)

Critical: In polyolefins, do not go beyond 240°C when EBS or surfactants are used as these components start to volatilize, resulting in odor and loss of performance.

Screw Speed:

- On twin-screw extruders, 80–120
- Use kneading blocks in zones 3–5 to improve mixing without undue shear heating.

Residence Time:

- At least 2.5 minutes. Less time may lead to incomplete wetting, especially in the case of carbon black or phthalocyanine-based pigments.

Vacuum:

- If vacuum is applied at the vent zone (if available), it will be effective in removing moisture as well as low-MW volatiles, this being most important in the cases of hygroscopic pigments or solvent-based masterbatches.

Pro Tip: Pre-dry pigments and masterbatch pellets to less than 0.05% moisture in the case of a high-speed film extrusion line. Water that is trapped in agglomerates leads to the formation of micro-bubbles and surface defects.

8.2 Filler Masterbatches

High filler loadings demand more than just mixing, they require true interfacial control. In mineral-filled compounds like HDPE or PP, poor wetting of $CaCO_3$, talc, or kaolin leads to agglomerates that act as stress concentrators, reduce impact strength, and clog filters. This section delivers real-world processing profiles that balance shear, temperature, and additive timing to achieve uniform dispersion, so you can run at high throughput without sacrificing mechanical performance or surface quality.

Target: Lower melt viscosity, avoid screen clogging, and

provide even distribution.

Temperature Profile:

- Feed Zone: 150–170°C (keep the area cool to avoid the settling of the filler or bridging)
- Melting Zone: 190–210°C (PE/PP); 200–220°C (PVC)
- Metering Zone: 200–220°C (PE/PP); 190–210°C (PVC), lower than the melting zone to prevent thermal degradation of the filler surface coatings

Screw Speed:

- On twin-screw extruders, 100–150
- Higher speeds result in higher shear, which, in turn, facilitates the breaking of agglomerates, however, this is only the case if the dispersant is present.
- Use reverse kneading elements in zones 4–5 to get high shear without too much heat buildup.

Residence Time:

- 3–4 minutes suggested. Mineral fillers need longer exposure time to be fully wetted.

Feeding Strategy:

- Side-feed the filler after the polymer has been melted (Zone 2 or 3). At the same time, feed a dispersant with the filler, never before.

Field Experience:

"If you put $CaCO_3$ into the feeder along with dry PE pellets, the screw will clog, and you'll have to clean it for about 4 hours. So, put it into the melt."

8.3 Engineering Plastics Compounding

Processing engineering plastics like PA, PBT, and PC demands more than just heat and shear, it requires precision chemistry. These resins are polar, high-performance systems where fillers, fibers, and flame retardants must be fully wetted to avoid agglomeration, fiber breakage, or dielectric failure.

This section delivers field-validated temperature profiles, screw designs, and additive strategies that ensure stable dispersion without degrading sensitive polymers. From moisture control to venting and residence time, it's all about balancing reactivity with robustness.

Target: Limit fiber breaking, improve flow, and keep phase separation from happening.

Temperature Profile:

- PA6/66: Feed 230°C → Melting 250–260°C → Metering 265–275°C
- PBT: Feed 220°C → Melting 240–250°C → Metering 255–265°C
- PC: Feed 260°C → Melting 280–290°C → Metering 285–300°C

  (only maximize temperature if the additive is thermally stable)

Screw Speed:

- 100–130 RPM. More shear = more fiber breakage. At the same time, low-shear mixing elements (e.g., Chevron or spiral elements) should be used for glass fiber reinforcement.

Residence Time:

- If less than 4 minutes is ensured. PBT and PC can hydrolyze and thermally degrade.

Moisture Control:

- Resins should be dried to less than 0.02% moisture prior to processing. If glass-fiber reinforcement is done with wet PA or PBT, voids will form and dispersion will be poor.

Vacuum:

- It is a must for PC and PBT to get rid of moisture and volatiles from additives.

Critical Insight: In the case of PBT with 20% glass fiber and 0.8% PE-g-MA dispersant, a 5°C reduction in metering temperature is responsible for a 15% decrease of melt viscosity and a 30% decrease of fiber breakage. It is advised not to overheat for "forcing flow" - rather chemistry should be used to achieve optimal flow first.

8.4 PVC Processing (Rigid & Flexible)

PVC isn't just another thermoplastic, it's a high-shear, low-temperature balancing act. In both rigid and flexible formulations, poor dispersion of stabilizers, pigments, or fillers leads to black specks, die lines, and early degradation. This section delivers field-proven processing guidelines that ensure uniform additive distribution without sacrificing thermal stability, because in PVC, a few degrees too hot can turn good mixing into costly scrap.

Target: Stabilizers and pigments to be dispersed

without degradation or die lines.

Temperature Profile:

- Feed Zone: 150–160°C
- Melting Zone: 170–180°C (very important to not get decomposition started early)
- Metering Zone: 180–190°C (do not go beyond 195°C)

Screw Speed:

- On single-screw extruders, 40–70
- High shear leads to the formation of localized hot spots, as a result of which, organotin or Ca/Zn stabilizers get degraded.

Residence Time:

- 3–5 minutes. PVC needs longer mixing for the full integration of stabilizers and lubricants.

Special Note:

- Plasticizer pre-mixes should be utilized for flexible PVC. For example, dispersants such as calcium stearate should be dry-blended with stabilizers before being added to resin.
- Remove silicone-based additives from rigid PVC if you want to co-extrude profiles that will laminate. They can cause the delamination.

8.5 Wood-Plastic Composites

Wood-plastic composites are only as strong as their weakest link, the interface between hydrophobic polyolefins and hydrophilic wood flour. Without proper coupling, moisture absorption, poor dispersion, and delamination lead to swelling, warpage, and premature failure. This section delivers field-proven strategies using functionalized polymers like PE-g-MA or PP-g-MA to chemically bond fibers to the matrix, reduce agglomeration, and improve mechanical performance. It's not about mixing wood into plastic; it's about making them work together.

Target: Achieve fiber wetting, reduce moisture absorption, and prevent die buildup.

Temperature Profile:

- Feed Zone: 140–150°C (keep low to prevent wood charring)
- Melting Zone: 170–180°C
- Metering Zone: 175–185°CWood degrades above 190°C, this is non-negotiable.

Screw Speed:

- 60–90 RPM. High shear tears fibers and increases surface area for moisture absorption.
- 60–90 RPM. High shear tears fibers and increases surface area for moisture absorption.

Residence Time:

- 4–5 minutes. Critical for PE-g-MA or PP-g-MA coupling agents to react with wood hydroxyl groups.
- 4–5 minutes. Critical for PE-g-MA or PP-g-MA coupling agents to react with wood hydroxyl groups.

Moisture Management:

- Dry wood flour to <2% moisture. Even 3% can cause steam bubbles and surface pitting.
- Dry wood flour to <2% moisture. Even 3% can cause steam bubbles and surface pitting.

Venting:

- Use dual vent ports to remove volatile organics and moisture generated during processing.
- Use dual vent ports to remove volatile organics and moisture generated during processing.

8.6 Rubber & TPE Compounding

Rubber and thermoplastic elastomers (TPEs) like SEBS, PP/EPDM, and TPO demand more than just pigment dispersion, they require additives that stabilize oil-based softeners, prevent filler agglomeration, and maintain long-term flexibility. This section delivers field-proven processing guidelines for systems prone to blooming, die build-up, and inconsistent cure behavior. You'll learn how to balance lubrication and compatibility without sacrificing surface quality or adhesion, especially in automotive seals and consumer grips where performance must last beyond the production line.

Target: Uniform dispersion of oil, fillers, and curatives without blooming or stickiness.

Temperature Profile:

- SEBS/TPO: Feed 160°C → Melting 180–190°C → Metering 190–200°C

- EPDM: Feed 150°C → Melting 170–180°C → Metering 180–190°C

Screw Speed:

- 50–80 RPM. Low shear preserves polymer chain integrity and prevents premature crosslinking.
- 50–80 RPM. Low shear preserves polymer chain integrity and prevents premature crosslinking.

Residence Time:

- 3–4 minutes. Longer times increase risk of oil migration or additive blooming.
- 3–4 minutes. Longer times increase risk of oil migration or additive blooming.

Additive Feeding Order:

- Feed oil and softeners after polymer melts. Feed dispersant (e.g., EBS or amide) with carbon black or silica.
- Feed oil and softeners after polymer melts. Feed dispersant (e.g., EBS or amide) with carbon black or silica.

Industry Standard:

A 0.3% addition of PE wax in TPE automotive grips reduces die build-up by 70%, improves surface finish, and does not change the hardness.

Table 8.6: General Best Practices Across All Applications

Parameter	Recommendation
Screw Design	Use 3-zone extruders with kneading blocks. Do not use long compression zones with high shear.
Filter Mesh	Begin with 80–100 mesh for masterbatches; 150–200 mesh for high-clarity films. Keep an eye on the pressure increase.

Melt Temperature Monitoring	Use inline melt pressure/temperature sensors, not just zone setpoints. The actual melt temp can be 10–20°C higher than the zone setting.
Startup Procedure	Always purge with a clean polymer before switching formulations. Residual dispersant can cause plate-out on new products.
Shutdown	Carry out 5–10 minutes of pure carrier resin (e.g., LDPE) to flush out additives and prevent degradation in the barrel.

These are not randomly chosen processing conditions, they represent the outcome of field experience, failure analysis, and production optimization over many years. It is not about running at maximum speed or temperature, but rather about establishing a stable, repeatable process where the additive works predictably, every time.

9. Dosages

When it comes to dosing a dispersing or a wetting additive, it is not a fixed number - it is a performance lever that is adjusted depending on the polymer chemistry, the nature of the additive, and the processing environment. Insufficient dosing will lead to incomplete dispersion while excessive dosing will cause plate-out, blooming, or mechanical degradation. Below are the dosage ranges that have been supported by evidence for the most common application along with the rationale, adjustment triggers, and industry benchmarks that help real-world formulation decisions.

9.1 Recommended Dosages

Getting dosage right isn't about following a recipe, it's about balancing performance and risk. Too little additive leads to poor dispersion, screen clogging, and specks; too much causes plate-out, blooming, or mechanical loss. This section provides field-validated dosage ranges for key applications, backed by KPIs and ROI from real production lines. It's not theory, it's what works when the batch is on the line and the customer is waiting.

Table 9.1: Recommended Dosages by Application (with Rationale & Guidance)

Application	Dosage Range (% by weight)	Rationale & Practical Guidance
Pigment Masterbatches (e.g., carbon black, phthalocyanine, titanium dioxide)	1.0 – 5.0%	High pigment loadings demand strong wetting and stabilization. Fine pigments (≤1 μm) require higher dosages (3–5%) to overcome agglomeration forces.

		For coarse pigments or low-color-strength systems (e.g., iron oxides), 1–2% may suffice. *Rule of thumb:* Increase dosage by 0.5% for every 10% increase in pigment concentration beyond 20%.
Filler Compounds ($CaCO_3$, talc, kaolin, mica)	0.5 – 2.0%	Mineral fillers are less cohesive than pigments but present high surface area. Lower dosages (0.5–1.0%) work well in HDPE/PP with low filler loadings (<30%). For high-load systems (40–60%), use 1.5–2.0%, especially with functionalized polymers like PE-g-MA. *Critical:* In pipe or profile extrusion, exceeding 2.0% can reduce stiffness and increase melt elasticity, causing die swell.
Engineering Plastics (PA, PBT, PC, PC/ABS)	0.5 – 3.0%	These systems often contain glass fiber, flame retardants, or impact modifiers, each demanding different wetting strategies. For glass fiber reinforcement, 0.5–1.5% functionalized polymer (e.g., PA-g-MA) is typical.

		For flame-retardant systems (e.g., brominated or phosphorus-based), dosages up to 3.0% may be needed to prevent bloom and ensure uniform distribution. *Caution:* In PC/ABS, overdosing silicone-based additives (>1.0%) can cause delamination in multi-layer parts.
PVC Systems (rigid and flexible)	0.5 – 1.5%	PVC's low thermal stability limits additive options. Metal soaps (e.g., calcium stearate) are typically used at 0.5–1.0%. For color masterbatches in rigid PVC, use 1.0–1.5% EBS or Montan wax, never exceed 1.5%, as higher levels accelerate degradation and cause yellowing. In flexible PVC, avoid silicone-based additives, they interfere with plasticizer migration and cause stickiness.
Rubber & TPE Compounds (SEBS, PP/EPDM, TPO)	0.3 – 1.5%	These systems often contain oil-based softeners and carbon black. Low dosages (0.3–0.8%) of amide-based dispersants (e.g., erucamide) improve pigment wetting and reduce die build-up. Higher dosages (1.0–1.5%) are used in high-filler TPOs (e.g.,

		30% talc) to maintain flow and surface finish. *Field Insight:* In automotive TPE grips, cut dosage from 1.2% to 0.7% and eliminated surface gloss variation, without losing slip.", TPE Formulator, Mexico
Wood-Plastic Composites (WPCs)	1.0 – 2.5%	The hydrophilic nature of wood flour demands strong coupling. PE-g-MA or PP-g-MA is typically used at 1.5–2.0%. Lower dosages (<1.0%) result in poor adhesion, leading to voids and moisture ingress. Higher dosages (>2.5%) increase melt viscosity excessively and can cause screw slippage. *Best Practice:* Pre-blend coupling agent with wood flour before adding to polymer to ensure direct contact.

Key Insight: "The dosage is not about what the supplier says, it's about what your extruder can handle and what your part requires. "Always confirm dosage with a filter pressure rise test and speck count before scaling up.

9.2 Overdosing Risks

It may appear that an allowed overdosing beyond the recommended dosages acts as a "safety margin," but it causes a series of significant, often concealed, problems:

Plate-out & Die Build-up:

Waxes- or amide-based additives migrate to metal surfaces (screw, die, screen) and accumulate, thus causing irregular flow, streaks, and unplanned shutdowns. They are common in PE/PP systems at levels higher than 2.5%.

Surface Blooming:

Additives move to the surface of the part with time, especially in the case of thin films or long-term outdoor applications, resulting in hazing, lower printability, or adhesive failure. Silicone and EBS are the main culprits in this respect.

Loss of Mechanical Properties:

An excess dispersant may serve as a plasticizer or a weak interface. In PA or PBT, >3.0% can cause the tensile strength to drop by 10–15%. In WPCs, an overdose of PE-g-MA makes the material more brittle.

Interference with Other Additives:

Overdosed dispersants can interfere with UV stabilizers, antioxidants, or flame retardants by competing for the interfacial sites.

Increased Cost & Waste:

Most of the high-performance additives (e.g., BYK-P 4100, Struktol TPW) are costly. Thus, a 0.5% overdose on 10 tons of a compound will result in an unnecessary cost of $500–$2,000.

9.3 Underdosing Consequences

Using too little dispersant might save on cost, until it costs you everything. Underdosing leads to incomplete pigment or filler dispersion, which means specks, streaks, and weak mechanical performance. It increases filter clogging, raises scrap rates, and can cause field failures that damage customer trust. This section shows why "just getting by" with low additive levels isn't cutting corners, it's creating them.

Under-dosing is just as expensive, if not more, because the problem is less visible and the product fails only in the field:

- Incomplete Dispersion: It causes specks, streaks, and color inconsistency, especially in black or dark masterbatches where even 0.1% of agglomerate is visible.
- Filter Clogging: Agglomerated fillers or pigments clog screen packs, which leads to increased pressure drop, output reduction, and forced shutdowns. In HDPE pipe extrusion, this can reduce line speed by 30%.
- Poor Mechanical Performance: In fiber-reinforced systems, underdosing causes the weakening of the fiber-matrix bond, thus leading to low impact strength and early failure.
- Inconsistent Processing: The system becomes unreliable for automated production due to high melt viscosity, high torque, and uneven flow.
- Customer Rejections: A batch containing specks in a consumer product can lead to a recall. In the automotive sector, the rejection of a single headlight housing can cost more than $10,000 in scrap and rework.

Pro Tip: It is advisable to start dosing at the lower end of the range. Incrementally increase the dose (0.2–0.3% at a time) while keeping an eye on:

- Melt pressure (extruder),
- Visual speck count (on molded plaques),

- Delta-E (color consistency),
- Filter pressure rise over 2 hours. The optimal dosage is the lowest amount that delivers consistent, defect-free performance.

These dosage guidelines are not theoretical; they are the distilled experience of compounders who have seen what happens when you get it wrong. Whether you are operating a lab extruder or a 24/7 production line, dosage is the most direct control you have over dispersion quality. Do not guess. Do not copy. Test, measure, and optimize.

10. Choosing the Right Dispersion and Wetting Additives

Identifying the best dispersing or wetting agent is not a matter of selecting either the "most potent" or "most pricey" product but rather a decision made at the system level by taking into account material compatibility, performance requirements, processing conditions, and cost-effectiveness. Choosing a wrong option here may result in production delays, rework that costs a lot, or even failures in the field. It is a comprehensive, practical framework that can be used to navigate your choice by merging all the previous technical insights into one workable flow.

10.1 Polymer Matrix Compatibility

Not all polymers are created equal, and neither are dispersants. What works in polyethylene will fail in polyamide. This section breaks down how resin polarity dictates additive selection: non-polar waxes for PE and PP, functionalized polymers for polar systems like PA or PET. It's not about preference; it's about chemical matching. Get it wrong, and you'll face poor dispersion, plate-out, or mechanical failure. Get it right, and you build performance from the interface up.

The chemical nature of the polymer determines the additives that will even non-polar resins (PE, PP, PS):

non-polar or weakly polar dispersants:

polyethylene wax, PP wax, EBS, or silicone masterbatches. They have molecular structures that "blend" with the matrix, thus ensuring good compatibility and minimal migration. Using surfactants or polar functionalized polymers should be avoided as these will phase-separate, bloom, or cause haze.

polar resins (PA, PBT, PC, PET, PVC):

Need polar or functionalized additives: surfactant-based systems (e.g., PEG esters), maleated polymers (PA-g-MA, PE-g-MA), or ester-based dispersants. These establish hydrogen bonds or dipole interactions with the resin, thus becoming firmly anchored at the interface.

Field Rule: "If the additive moves to the top or after cooling it creates a greasy film, it is not compatible. Changing the additive before running 10 tons is the wise decision.", Masterbatch Plant Manager, Italy

10.2 Additive-Pigment/Filler Affinity

Not all additives stick to all particles, chemistry matters. This section breaks down how dispersants bond (or fail to bond) with different pigments and fillers based on surface energy and polarity. You'll learn why PE wax works with $CaCO_3$ but not carbon black, why silicone excels with nano-fillers, and when functionalized polymers like PE-g-MA are non-negotiable. It's not about adding more additive, it's about choosing one that actually sticks.

The target solid must be *wetted* before it can be *dispersed*. This is determined by chemistry:

Table 10.2: Additive/Pigment/Filler Affinity Matrix

Additive Type	Preferred Dispersant Chemistry	Why
Carbon Black / Organic Pigments	Functionalized polymers (PE-g-MA), silicone masterbatches	High surface energy; it requires strong adsorption and steric shielding.
TiO_2 / $CaCO_3$ / Talc	Polyolefin waxes, EBS, metal soaps	Medium surface energy; low-cost waxes are most of the time enough.
Glass / Carbon Fibers	Maleated polyolefins (PP-g-MA, PA-g-MA)	Hydrophilic fibers that need coupling to hydrophobic matrix.

Clay / Silica (Nano-fillers)	Surfactant-based or silane-modified dispersants	High length/diameter ratio and agglomeration tendency require steric stabilization.

Pro Tip: Check affinity by taking a small amount of additive and mixing it with the filler in a vial. If the mixture slurry or paste, then they are compatible. In case it clumps or separates, it will not work in the melt.

10.3 End-Use Requirements

The final application defines what "good" dispersion really means. A masterbatch that passes visual inspection may still fail in sterilization, migration testing, or outdoor exposure. This section cuts through generic advice to focus on performance-critical requirements: FDA/EFSA compliance for food contact, long-term gloss and clarity retention, and resistance to UV degradation. You'll learn how additive choice directly impacts regulatory approval, shelf life, and field reliability, because if it doesn't perform after processing, it doesn't matter how well it mixed.

An additive of yours has to meet the demands of the final product both regulatory and functional, no exceptions are allowed.

- Food Contact (FDA, EU 10/2011): Only add those additives that are listed in the regulatory databases. Do not use heavy metals (e.g., zinc stearate in some applications), phthalates, or low molecular weight surfactants that may migrate. Licowax PE 520 and Struktol TPW 100 are FDA-compliant for polyolefins. It is Always recommended to receive a Letter of Guarantee (LOG) from the supplier.
- Clarity / Optical Performance: In transparent films or lenses, even 0.1% of an incompatible dispersant can cause haze. Use low-migration, high-purity agents: silicone masterbatches (e.g., Silastic™ MB) or specially formulated polyolefin waxes. Do not use amide-based additives (erucamide) in clear PP, there is a high risk of bloom.
- UV Stability / Outdoor Exposure: Refrain from using additives that are susceptible to oxidation or photodegradation (e.g., some fatty acid

derivatives). Use thermally stable, non-volatile agents such as Montan wax or functionalized polymers. In WPCs or outdoor decking, make sure that the dispersant is not the one that is causing the wood component to be aged faster.

10.4 Cost vs. Performance Trade-offs

Choosing a dispersing additive isn't about minimizing cost, it's about maximizing value. In industrial compounding, the cheapest option often carries hidden expenses: scrap, downtime, rework, and customer rejections. This section helps you weigh upfront additive cost against total cost of ownership, so you invest wisely, whether you're making a commodity film or a medical-grade component.

Within the industry, a raise in performance is never without a cost, but neither is an instance of failure.

Table 10.4: Cost vs. Performance Trade-offs by Scenario

Scenario	Recommended Approach
High-volume, low-margin product (e.g., HDPE trash bags, basic packaging)	Use EBS or PE wax. Dispersion quality can be at a mediocre level for the sake of saving money. Process stability should be given more importance than the cosmetic perfection.
High performance, regulated product (e.g., automotive interior, medical device, food packaging)	Put your money into functionalized polymers (PE-g-MA) or specialty surfactants (BYK-P 4100). The higher price is paid off by the lower volume of scrap, fewer customer complaints, and brand protection.
New formulation, unknown performance	Implement a blended strategy: 0.5% PE wax + 0.3% PE-g-MA. With this setup, you get the cost advantage of wax and the coupling strength of the functional polymer. Test, measure, and then optimize.
Overdosing to "fix" poor dispersion	Do not overdosing to "fix" poor dispersion. It is a band-aid solution. Instead, look at

your additive type, dosage, and processing method once again. Most of the time, the root cause is not "not enough".

Cost Calculation Example:

By adding 1.0% of $25/kg BYK-P 4100 to a 10-ton batch = $2,500. However, if that additive is the cause of a 5% scrap elimination in a $1.20/kg product, then the batch saves $60,000.

The right additive does not cost more but rather pays for itself.

10.5 Decision Framework: The 5-Question Checklist

Choosing the right dispersant doesn't have to be guesswork. This section distills everything, from resin polarity to end-use requirements, into five practical questions you can ask on the spot. It's a field-tested checklist used by top compounders to eliminate trial-and-error and make confident, data-driven decisions. If you're balancing performance, cost, and compliance under production pressure, this is your go-to filter before scaling any new formulation.

Before you making your final decision on an additive, consider these five questions, one after another:

1. What is the polymer matrix?

→ *Non-polar (PE/PP)? Polar (PA/PBT)?*

2. What are you dispersing?

→ *Fine pigment? High-load mineral? Glass fiber?*

3. What are the end-use requirements?

→ *Food contact? Clarity? UV resistance?*

4. What's my processing setup?

→ *Twin-screw extruder? Single-screw? High shear? Low residence time?*

5. What's my acceptable cost per kg?

→ *Is this a commodity or premium product?*

Answer these questions honestly, and the choice of your additive is not a matter of speculation but clear.

This part of your guide changes the reference manual into a decision engine. It not only informs you *what* to use but also *why*, *when* and how to keep away from the costly mistakes that production derail.

11. Use Cases

Real-world dispersion challenges don't appear in textbooks, they appear in production lines at 3 a.m., with a rejected batch on the floor and a customer breathing down your neck. Below are three documented case studies from actual compounding operations, each demonstrating how the right dispersing strategy turned a costly failure into a reliable, scalable process.

11.1 Case 1: Improving $CaCO_3$ dispersion in HDPE pipe compound

Problem:

A pipe extruder of PE100 grade 110mm HDPE pressure pipe with 45% $CaCO_3$ filler content was experiencing frequent screen pack clogging (every 4–6 hours), inconsistent surface finish, and out-of-spec hoop stress values. White specks and streaks on the pipe surface were detected by the naked eye. Scrap rate had increased from 5% to 18% of the formulation.

Root Cause Analysis:

The existing formulation utilized 1.0% EBS as the main dispersant/lubricant. Although EBS facilitated flow, it did not provide stabilization, $CaCO_3$ particles were even though they were initially re-agglomerated during cooling. The twin-screw extruder (35mm, L/D = 40) was running at 110 RPM with a melt temperature of 205°C, but the feed zone was too hot (180°C), thus causing premature melting and poor filler wetting.

Solution:

- Substituted EBS with 1.8% PE-g-MA (Polyethylene-graft-maleic anhydride) as the main dispersant/coupling agent.
- Added 0.2% Montan wax (for controlled lubrication and melt stability).
- Temperature profile was adjusted:
    - Feed Zone: 150°C → 140°C (to prevent premature softening of filler)

- Melting Zone: 190°C → 195°C
- Metering Zone: 205°C → 200°C (reduced to avoid thermal degradation of PE-g-MA)

- The kneading block was installed in Zone 4 to increase shear without the screw speed.

Results (After 30 Days of Production)

Table 11.1: Improving $CaCO_3$ Dispersion in HDPE Pipe and Results

Metric	Before	After	Improvement
Screen Change Frequency	Every 4.5 hours	Every 24+ hours	+433%
Filter Pressure Rise (ΔP)	+28 bar over 2h	+4 bar over 2h	86% reduction
Visual Speck Count (per 10 cm²)	12–18	0–2	>85% reduction
Hoop Stress (ASTM D1599)	8.2 MPa (below spec: 9.0 MPa)	9.7 MPa	Met spec with margin
Scrap Rate	18%	4.2%	77% reduction

Cost Impact:

- PE-g-MA cost: $22/kg vs. EBS at $8/kg → +$14/kg additive cost
- Savings: $120,000/month in scrap reduction + 12 fewer shutdowns/month → ROI in 11 days

Lesson Learned:

"EBS is a lubricant, it doesn't bond filler to polymer. PE-g-MA isn't just a dispersant; it's a molecular weld. We stopped trying to lubricate our way out of dispersion problems.", Lead Compounder, Middle East Pipe Manufacturer.

11.2 Case 2: Eliminating specks in black masterbatch for automotive PP

Problem:

An automotive supplier producing black PP masterbatches (30% carbon black in PP) was rejecting 12–15% of batches due to visible specks in injection-molded interior trim parts. The specks were not pigment agglomerates; they were *unwetted carbon black clusters* that survived extrusion and appeared as dark dots on light-colored surfaces.

Root Cause Analysis:

This masterbatch was produced on a 25mm twin-screw extruder at 120 RPM, using 2.5% EBS and 0.5% PE wax. Carbon black (CB) had a surface area of 800 m^2/g, which is extremely high. The additive package lacked sufficient steric stabilization, and the melt residence time was only 1.8 minutes. Visual inspection confirmed incomplete wetting under microscope: 30–50% of CB particles were still clustered.

Solution:

- BYK-P 4100 1.2% (a high-performance, polar-compatible dispersant for carbon black) was used to replace EBS and PE wax.
- The total dosage was reduced to 1.7% (from 3.0%) by removing the redundant additives.
- The residence time was extended to 3.2 minutes by decreasing the screw speed to 95 RPM and lengthening the mixing section by 20cm.
- Carbon black was pre-dried to less than 0.03% moisture (using a vacuum dryer at 80°C for 4 hours).
- Inline melt temperature monitoring was implemented, and the actual melt temperature was confirmed to be 228°C, not 215°C as set.

Results (After 45 Days of Production):

Table 11.2: Eliminating Specks in Black Masterbatch and Results

Metric	Before	After	Improvement
Speck Count (ISO 1133-1, 10x magnification)	18–25 specks/10 cm²	0–2 specks/10 cm²	>90% reduction
Color Strength (ΔE ab)	ΔE = 2.8 (out of spec: <1.5)	ΔE = 0.9	Met spec
Melt Flow Index (MFI)	14.2 g/10min	14.8 g/10min	Improved flow
Batch Rejection Rate	14%	1.1%	92% reduction

Cost Impact:

- BYK-P 4100 cost: $45/kg → +$28/kg vs. previous blend
- Savings: $98,000/month in reduced scrap + 18 fewer customer complaints/month → ROI in 8 days

11.3 Case 3: Enhancing carbon fiber distribution in PBT for electrical connectors

Problem:

A manufacturer of PBT (20% glass fiber, 5% carbon fiber) for high-voltage electrical connectors had inconsistent dielectric strength and warpage issues in the molded parts. Electron microscopy showed that carbon fiber was clumping along the mold flow lines, thus creating localized conductive paths. Parts failed 15% of the electrical leakage tests.

Root Cause Analysis:

The compound contained 1.0% EBS and 0.8% PE wax. Carbon fiber (CF) is hydrophilic, on the other hand, PBT is polar but not reactive. The dispersant did not provide coupling, CF fibers were physically mixed but not chemically bonded, resulting in poor interfacial adhesion and fiber pull-out.

Solution:

Substituted EBS/PE wax with 1.5% PA-g-MA (Polyamide-graft-maleic anhydride), selected for its ability to Interact with both PBT's ester groups and CF's surface oxides.

- Reduced total additive dosage to 1.7% (from 1.8%) by removing the incompatible waxes.
- Adjusted extruder profile:
  - Feed Zone: 220°C → 215°C
  - Melting Zone: 250°C → 245°C (to prevent PBT degradation)
  - Metering Zone: 260°C → 255°C
  - Installed reverse kneading elements in Zones 3–4 to gently align fibers without breaking them.
  - Carried out moisture control: PBT was dried to 0.015% before compounding.

Results (After 60 Days of Production):

Table 11.3: Enhancing carbon fiber distribution in PBT for electrical connectors

Metric	Before	After	Improvement
Fiber Distribution (SEM imaging)	40% agglomerates	<5% agglomerates	87% improvement
Dielectric Leakage (IEC 60112)	18% failure rate	1.2% failure rate	93% reduction
Warpage (flatness tolerance)	±0.8 mm	±0.2 mm	75% improvement
Tensile Strength (ASTM D638)	102 MPa	118 MPa	+16% increase

Cost Impact:

- PA-g-MA cost: \$38/kg → +\$25/kg vs. previous blend
- Savings: Avoided \$220,000 in warranty claims from field failures + 22% reduction in rework → ROI in 14 days

Lesson Learned:

"You can't 'mix' carbon fiber into PBT, you have to chemically marry it. EBS is for lubrication. This is structural engineering. We didn't fix dispersion, we fixed the interface.", Materials Engineer, Automotive Electronics, USA

11.4 Case 4: Preventing Plate-Out in High-Gloss PP Automotive Trim with Carbon Black

Problem:

A supplier producing high-gloss black PP interior trim (dashboards, door handles) was experiencing recurring plate-out on mold surfaces, a sticky, black residue that built up every 8–12 hours, requiring manual cleaning and causing surface defects (haze, loss of gloss). Scrap rates climbed to 10%, and downtime cost ~\$18,000/day.

The masterbatch used 2.0% EBS and 0.5% PE wax for dispersion. While color was uniform, the low-molecular-weight additives migrated to the mold surface during injection molding, especially under high cavity pressure (1200 bar) and elevated mold temperatures (85°C).

Root Cause:

EBS and PE wax have low molecular weight and high mobility, ideal for dispersion but terrible for long-term stability in high-heat, high-pressure environments. They were not chemically anchored and simply migrated out of the matrix.

Solution:

- Replaced EBS/PE wax with 1.0% BYK-P 4100 (a high-MW, polar dispersant designed for carbon black in PP).

- Added 0.3% high-MW polyolefin-based stabilizer (Struktol TPW 150) to suppress migration without sacrificing dispersion.
- Reduced total dosage from 2.5% to 1.3%.
- Adjusted injection parameters:
  - Mold Temperature: 85°C → 75°C (to reduce additive mobility)
  - Holding Pressure: 1200 bar → 1000 bar (reduced stress-induced migration)
  - Cycle Time: 45 sec → 52 sec (slightly longer cooling to lock in dispersion)

Results (After 60 Days):

Table 11.4: Preventing Plate-Out in PP Trim and Results

Metric	Before	After	Improvement
Plate-Out Frequency	Every 9 hours	Every 72+ hours	+700%
Gloss (60° GI)	78 GU	89 GU	+14% increase
Surface Defects (specks + haze)	12 per 100 parts	1 per 100 parts	92% reduction
Scrap Rate	10%	1.5%	85% reduction
Cleaning Downtime	4 hrs/week	0.5 hrs/week	87% reduction

Cost Impact:

- BYK-P 4100 + Struktol TPW 150 cost: ~$32/kg vs. $8/kg for EBS/PE wax → +$24/kg
- Savings: $145,000/month in reduced downtime, scrap, and labor → ROI in 11 days

Lesson Learned:

"We thought we needed more lubricant to fix gloss. We actually needed less migration. The right dispersant doesn't just mix, it stays put.", Process Engineer, Automotive Tier-1, Germany

11.5 Case 5: Solving Color Bleeding in PE Film for Food Packaging

Problem:

A flexible packaging producer using red pigment (azo-based) in LDPE film for snack bags was receiving customer complaints about color bleeding onto adjacent white layers during heat sealing. The seal strength also dropped by 18% in accelerated aging tests.

The pigment was compounded at 0.8% using 1.2% EBS. While initial color was strong, the pigment migrated to the film surface during storage (30–40°C) and bled into the sealant layer during heat sealing (150°C).

Root Cause:

EBS is a migration-prone external lubricant. It acted as a carrier for the pigment, pulling it to the surface over time. The pigment itself had poor affinity for LDPE and no steric shielding.

Solution:

- Replaced EBS with 0.7% Hydrocerol® (Clariant's wax-based dispersant specifically engineered for food-contact color masterbatches).
- Added 0.2% Polyethylene glycol (PEG) ester (non-migrating, FDA-compliant) to enhance pigment anchoring.
- Reduced total dosage from 2.0% to 0.9%.
- Implemented post-extrusion quenching: Film passed through a chilled water bath (15°C) immediately after die exit to "freeze" the dispersion.

Results (After 90 Days of Production):

Table 11.5: Solving Color Bleeding in PE Film and Results

Metric	Before	After	Improvement
Color Bleeding (visual on seal)	15% of rolls affected	0% affected	100% eliminated
Seal Strength (ASTM F88)	1.8 N/mm	2.1 N/mm	+17% increase
Migration Test (FDA 21 CFR 178.3820)	0.8 ppm pigment in sealant	<0.05 ppm	94% reduction
Shelf-Life Stability (60°C, 14 days)	Color shift ΔE = 3.2	ΔE = 0.7	Met food contact spec
Customer Complaints	12/month	0	100% eliminated

Cost Impact:

- Hydrocerol® + PEG ester cost: ~$28/kg vs. EBS at $8/kg → +$20/kg
- Savings: Avoided $80,000 in customer penalties + saved 300+ hours of rework/month → ROI in 14 days

Lesson Learned:

"In food packaging, you're not just making a film, you're making a safety barrier. If the pigment can move, it's not dispersed, it's contaminated.", Quality Manager, Food Packaging, USA

11.6 Case 6: Reducing Die Build-Up in TPE Door Seals with High Silica Loading

Problem:

A manufacturer producing TPE (SEBS-based) door seals for automotive applications was facing severe die build-up in extrusion lines producing 15mm profiles. Silica filler (25%) was causing sticky residue on die lips every 6–8 hours, leading to inconsistent profile dimensions and surface roughness.

The formulation used 1.5% erucamide as a slip agent and 0.8% EBS for dispersion. Silica's surface silanol groups were attracting polymer chains and forming a gel-like layer on the die.

Root Cause:

Erucamide and EBS were migrating to the die surface, where they reacted with silica's hydroxyl groups, forming a tacky, crosslinked residue. The additives were acting as *adhesives*, not dispersants.

Solution:

- Replaced erucamide and EBS with 1.0% Silastic™ Additive MB-100 (Dow's silicone masterbatch designed for silica-filled TPEs).
- Added 0.2% titanate coupling agent (Tyzor® TPT) to chemically bond silica to the SEBS matrix.
- Reduced total additive load from 2.3% to 1.2%.
- Modified die geometry:
  - Polished die land to mirror finish (Ra < 0.2 μm)
  - Increased die land length by 15% to stabilize flow

Results (After 45 Days):

Table 11.6: Reducing Die Build-Up in TPE Door Seals and Results

Metric	Before	After	Improvement
Die Build-Up Frequency	Every 7 hours	Every 140+ hours	+1900%
Profile Dimension Tolerance	±0.3 mm	±0.08 mm	73% improvement
Surface Roughness (Ra)	3.8 μm	1.1 μm	71% reduction

Extrusion Throughput	18 kg/hr	22 kg/hr	+22% increase
Maintenance Cost	$2,200/month	$300/month	86% reduction

Cost Impact:

- Silastic™ MB-100 + Tyzor® cost: ~$40/kg vs. $8/kg for erucamide/EBS → +$32/kg
- Savings: $110,000/year in labor, downtime, and die rework → ROI in 18 days

Lesson Learned:

"Silica isn't just a filler; it's a chemical trap. You can't lubricate your way out of this. You need chemistry that silences the surface.", Extrusion Manager, Automotive Seals, Mexico

11.7 Case 7: Eliminating Color Shift in White PP Injection Molding for Medical Devices

Problem:

A medical device manufacturer producing white PP components (syringe housings, IV connectors) was experiencing yellowing and inconsistent whiteness after sterilization (ethylene oxide and gamma irradiation). Batch-to-batch Delta-E values varied between 2.1 and 4.8, well outside the acceptable limit of ≤1.5. The issue was traced to titanium dioxide (TiO_2) agglomeration, which created localized hot spots during sterilization, accelerating oxidative degradation.

The formulation used 3.0% rutile TiO_2 with 0.8% EBS as the dispersant. While the parts looked white off the mold, post-sterilization color drift was severe.

Root Cause:

EBS provided no long-term stabilization. TiO_2 particles were wetted initially but re-agglomerated during cooling, creating clusters with higher thermal mass. During sterilization, these clusters absorbed and retained heat more than the surrounding polymer, triggering localized photo-oxidation and yellowing.

Solution:

- Replaced EBS with 1.2% BYK-P 4100 (a high-performance, polar dispersant proven for TiO_2 in polyolefins).
- Added 0.3% hindered amine light stabilizer (HALS) specifically formulated for polyolefins to suppress radical degradation.
- Reduced total additive load from 3.8% to 4.5% (slight increase due to HALS, but *dispersant dosage dropped from 0.8% to 1.2%*, more effective chemistry).
- Implemented controlled cooling: Mold temperature reduced from 40°C to 25°C, with forced air impingement to rapidly lock in dispersion.

Results (After 90 Days of Production):

Table 11.7: Eliminating Color Shift in White PP Medical Parts and Results

Metric	Before	After	Improvement
Delta-E Post-Sterilization	2.1 – 4.8	0.6 – 1.2	>90% reduction in variation
Whiteness Index (ASTM E313)	82.5	88.3	+7% improvement
Reject Rate (Color)	14%	1.8%	87% reduction
Sterilization Cycle Time	4 hrs	4 hrs (unchanged)	No change needed
Customer Audit Pass Rate	72%	100%	Zero non-conformances

Cost Impact:

- BYK-P 4100 + HALS cost: ~$35/kg vs. EBS at $8/kg → +$27/kg
- Savings: Avoided $190,000/year in rework, recalls, and audit failures → ROI in 10 days

Lesson Learned:

"You can't fix color stability with more pigment. You fix it by ensuring every TiO_2 particle is alone, and protected. Dispersion isn't about color, it's about molecular isolation.", R&D Lead, Medical Plastics, Switzerland

11.8 Case 8: Stabilizing Flame Retardant Dispersion in ABS for Telecom Housings

Problem:

A producer of ABS telecom enclosures (routers, base stations) was failing UL94 V-0 flammability tests inconsistently. Some batches passed, others didn't, despite identical formulations. Inspection revealed clumping of brominated flame retardant (BFR), which created localized areas with insufficient flame inhibition.

The compound used 20% BFR with 1.0% EBS and 0.5% PE wax. The BFR had a high melting point (210°C) and low surface energy, making it notoriously difficult to wet in ABS.

Root Cause:

EBS and PE wax were incompatible with the polar BFR molecules. The dispersants coated the ABS matrix but failed to penetrate or stabilize the BFR particles. During extrusion, BFR remained in large, unwetted clusters, leading to uneven distribution and failure under flame testing.

Solution:

- Replaced EBS/PE wax with 1.5% Struktol® TPW 100 (a multi-functional dispersant designed for halogenated flame retardants in engineering thermoplastics).

- Added 0.2% phosphorus-based synergist (to reduce BFR loading by 1% without sacrificing rating).
- Reduced BFR from 20% to 19%, lowering melt viscosity and improving processability.
- Adjusted extruder profile:
  - Feed Zone: 200°C → 210°C (to pre-soften BFR)
  - Melting Zone: 240°C → 250°C
  - Metering Zone: 245°C → 240°C (to avoid thermal degradation of ABS)
- Used reverse kneading elements in Zones 4–5 to break up BFR clusters without shear degradation.

Results (After 60 Days):

Table 11.8: Stabilizing Flame Retardant Dispersion in ABS and Results

Metric	Before	After	Improvement
UL94 V-0 Pass Rate	68%	100%	+32 percentage points
Melt Flow Index (MFI)	12.5 g/10min	15.2 g/10min	+22% increase
BFR Clustering (SEM)	40% of particles agglomerated	<5% agglomerated	87% reduction
Surface Appearance	Slight specking	Smooth, uniform	Eliminated
Raw Material Cost	$4.20/kg	$4.05/kg	3.6% reduction (due to lower BFR load)

Cost Impact:

- Struktol® TPW 100 cost: $38/kg → +$30/kg vs. EBS/PE wax
- Savings: $280,000/year from eliminated test failures, rework, and reduced BFR usage → ROI in 6 days

Lesson Learned:

"Flame retardants don't need to be mixed; they need to be integrated. If your dispersant can't touch the BFR, your part will burn. We stopped trying to lubricate our way to safety.", Fire Safety Engineer, Telecom Manufacturing, Taiwan

11.9 Cross-Case Insight: The Universal Laws of Dispersion in Industrial Plastic Processing

Across all eight real-world cases, from HDPE pipe to medical syringes, automotive trim to telecom housings, a clear, consistent pattern emerges. These are not isolated fixes. They are universal truths about how dispersion *actually* works under industrial pressure. What separates success from costly failure is not equipment, not speed, not even dosage, it's understanding the interface.

11.9.1 1. EBS and Low-MW Waxes Are Not Dispensers, They Are Migrators

In 7 out of 8 cases, EBS, PE wax, or erucamide were the *original* dispersants, and the *root cause* of the failure.

- They provided temporary flow improvement but no long-term stabilization.
- They migrated to surfaces during processing (plate-out in PP trim), storage (color bleeding in film), or sterilization (yellowing in medical parts).
- In Case 6, they didn't just migrate, they *reacted* with silica, forming a glue-like residue.

Truth: EBS is a lubricant. It helps the melt flow. But it does *not* anchor particles. When you need dispersion, you need *anchoring*, not slip.

11.9.2 The Right Additive Is Not the Cheapest, It's the One That Prevents the Most Expensive Failure

Every successful case involved a higher-cost additive, BYK-P 4100 ($45/kg), PA-g-MA ($38/kg), Silastic™ MB-100 ($40/kg), yet all achieved ROI in under 2 weeks.

- Case 8: $30/kg more additive → $280,000/year saved by eliminating UL94 failures.
- Case 5: $20/kg more → $80,000/year saved by preventing customer penalties.
- Case 1: $14/kg more → $120,000/month saved by cutting scrap and downtime.

Truth: You don't save money by choosing cheap additives. You save money by choosing *effective* ones. The real cost is not the price per kg, it's the cost per rejected batch, per recall, per shutdown.

11.9.3 Dosage Reduction = Performance Gain (When Chemistry Is Right)

In every case, the *total additive load was reduced* after switching to targeted chemistry:

- Case 2: 3.0% → 1.7%
- Case 4: 2.5% → 1.3%
- Case 5: 2.0% → 0.9%
- Case 6: 2.3% → 1.2%
- Case 8: 1.5% → 1.5% (but replaced ineffective + added synergist, net improvement)

This wasn't coincidence. High dosages masked poor compatibility. The new additives were more efficient per molecule; they worked *better* with *less*.

Truth: Dispersion is not a quantity problem. It's a quality problem. Less is more, when chemistry is precise.

11.9.4 The Real Problem Is Not Dispersion, It's Stabilization

Wetting is easy. Dispersion is hard. Stabilization is the invisible battle.

- In Case 1 ($CaCO_3$): Particles were wetted but re-agglomerated during cooling → solved by PE-g-MA's steric shielding.
- In Case 7 (TiO_2): Particles were mixed but clustered under heat → solved by BYK-P 4100's molecular isolation.
- In Case 3 (Carbon Fiber): Fibers were dispersed but pulled out → solved by PA-g-MA's covalent-like coupling.

Truth: If your additive doesn't *keep* particles apart after processing, you haven't solved dispersion, you've just delayed failure.

11.9.5 End-Use Requirements Dictate Chemistry, Not Vice Versa

The selection criteria were never "what's available." They were:

- Medical devices (Case 7): Must pass sterilization → required HALS + non-migrating dispersant.
- Food packaging (Case 5): Must meet FDA 21 CFR 178.3820 → required Hydrocerol® and PEG ester.
- Electrical connectors (Case 3): Must pass IEC 60112 → required coupling, not just dispersion.
- Automotive trim (Case 4): Must maintain gloss for 10+ years → required high-MW, non-migrating agents.

Truth: You don't choose an additive because it's "good for polyolefins." You choose it because it's the only one that survives your product's real-life environment.

11.9.6 Process Adjustments Follow Chemistry, Never the Other Way Around

In every case, equipment changes came *after* the additive was fixed:

- Screw speed lowered (Case 2, 3, 8)
- Mold temperature reduced (Case 4, 7)
- Cooling accelerated (Case 5, 7)
- Die land polished (Case 6)
- Residence time extended (Case 2, 3)

These weren't workarounds, they were optimizations enabled by chemistry. You don't fix a bad dispersion with more shear. You fix it with the right chemistry, *then* you tune the machine.

Truth: Don't try to force your machine to fix your formulation. Fix your formulation, then let your machine shine.

11.9.7 "Compatibility" Is Not a Label, It's a Functional Relationship

The term "compatible with PP" is meaningless. What matters is:

- Does it adsorb onto the particle?
- Does it anchor to the polymer?
- Does it survive the processing and end-use environment?

In Case 8, EBS was "compatible" with ABS, but failed to interact with BFR.

In Case 6, EBS was "compatible" with SEBS, but reacted with silica.

In Case 5, EBS was "compatible" with LDPE, but carried pigment to the surface.

Truth: Compatibility is not chemical similarity, it's functional synergy. Ask: *Does it do its job without causing collateral damage?*

11.9.8 The Best Dispersion Is Invisible, Until It Fails

No one notices a perfectly dispersed masterbatch.

But everyone notices:

- A speck on a dashboard
- A leaky electrical connector
- A discolored syringe
- A sticky die that shuts down production

The best dispersant doesn't make your compound look better.
It makes your production reliable, your product safe, and your brand trusted.

Final Truth:

- Dispersion isn't a chemical step. It's a risk mitigation strategy.
- You're not adding an additive to improve mixing.
- You're adding an additive to prevent a recall.
- To avoid a shutdown.
- To meet a regulation.
- To keep your customer, and your job.

The Dispersion Mindset: A Practical Checklist for Formulators

Before you finalize any formulation, ask these five questions:

1. Is this additive *anchoring* or *migrating*?
   → If it moves, it's not solving the problem, it's creating one.

2. Does this dosage reflect *efficiency* or *compensation*?
   → High dosage = masking poor chemistry. Low dosage + good chemistry = excellence.
3. Will this additive survive my *end-use environment*, not just my extruder?
   → UV, heat, moisture, sterilization, food contact, they all demand different chemistry.
4. Am I using this additive to *fix* a problem, or to *prevent* one?
   → Reactive fixes cost more. Proactive chemistry saves more.
5. Would I bet my reputation on this formulation?
   → If the answer isn't a confident "yes," go back to the lab.

These eight cases prove one thing:

The best dispersion isn't found in a catalog; it's engineered at the interface.

Choose your additive like you choose your safety gear:

- Not by price.
- Not by habit.
- But by performance under pressure.

Because in industrial plastics, when dispersion fails, it's not a batch that's lost.
It's your credibility.

12. Regulatory Restrictions and Limitations

One of the main features of industrial plastic processing is that regulatory compliance is a non-negotiable pillar of product integrity and not an afterthought. A lab-dispersion or wetting additive that looks great on the performance side can make a whole food production line non-compliant if the product violates food contact, environment, or emissions regulations. If one disregards these limitations, they expose themselves and their company to recalls, fines, and the irreversible losing of brand trust. The three crucial regulatory areas that need to be inspected at the formulation stage are listed below.

12.1 Food Contact Compliance

Just because it mixes well doesn't mean it's safe to eat from. This section cuts through the confusion around food contact regulations, focusing on what actually matters: migration limits, acceptable additives, and verifiable documentation. You'll learn which dispersants are approved under FDA 21 CFR 178.3820 and EU Regulation 10/2011, why EBS and low-MW waxes often fail in sensitive applications, and how to validate compliance with Letters of Guarantee (LOG), not guesswork. It's not about avoiding risk, it's about proving safety, batch after batch.

In packaging food, beverage packaging, medical devices, or infant products, additive migration to food or simulants is strictly regulated.

- FDA 21 CFR 178.3820 (U.S.) and EU Regulation 10/2011 specify migration limits (SMLs) for substances that may be transferred into food. A lot of typical dispersants, especially low-molecular-weight waxes, surfactants, and amide-based lubricants (e.g., EBS, erucamide), are either not listed or have very low SML thresholds (e.g., 0.05 ppm for some pigments in Case 5).

- Acceptable Choices:
  - Hydrocerol® (Clariant): Explicitly approved for food contact in polyolefins.
  - Licowax PE 520: It is a component that is listed under FDA 178.3820 for use in PE and PP.
  - Struktol® TPW 100/150: Food contact certified in tightly defined applications (check with supplier LOG).
- Critical Pitfalls:
  - If EBS or oleamide are used in direct food contact films even if "technically compatible" they can induce regulatory non-conformances resulting from migration.
  - Silicone-based additives may be compliant in certain applications but generally restricted in high-temperature food processing due to the possibility of volatilization.
  - Action Step: Always obtain a Letter of Guarantee (LOG) from your additive supplier that indicates the exact grade, the intended polymer, and the regulatory compliance status. Do not assume that a product is compliant just because it is a "general use" claim.

12.2 REACH & RoHS Considerations

Compliance isn't just about performance; it's about market access. In the EU and global supply chains, using an additive that violates REACH or RoHS can shut down production, block exports, or void customer contracts. This section cuts through the regulatory noise to focus on what matters: identifying SVHCs (Substances of Very High Concern), avoiding restricted additives like certain phthalates or heavy metals, and verifying supplier documentation. It's not enough to assume compliance, top compounders demand proof. Because in real-world manufacturing, a missed restriction doesn't just fail in testing. It fails in customs, in audits, and on the customer's floor.

A part from food safety, regulations respecting the environment and human health govern the use of chemicals in the EU and global markets.

- ·REACH (Registration, Evaluation, Authorization and Restriction of Chemicals) demands that full disclosure is made of Substances of Very High Concern (SVHC). Several old dispersants may either contain or be derived from:
  - Phthalates (e.g., in some plasticizer-based wetting agents), restricted under REACH Annex XIV.
  - Heavy metals (e.g., lead, cadmium in some pigment carriers), completely banned.
  - Certain surfactants with PBT/vPvB properties (persistent, bioaccumulative, toxic).
  - RoHS (Restriction of Hazardous Substances) is related to electronics and electrical equipment. The main targets of RoHS are Pb, Hg, and BFRs (lead, mercury, and brominated flame retardants), but it indirectly impacts dispersants that are used in ABS telecom housings (Case 8) or PBT connectors, where FBRs and their carriers must be carefully selected.
- Best Practice:
  - Get the REACH-compliant products from BYK, Clariant, or Schill + Seilacher and their suppliers.
  - Do not use generic "EBS" from non-certified sources, several of them may have undeclared impurities in that raw material, which are not mentioned in the SDS.
  - Maintain a List of Substances in your technical file, that is updated regularly, and tracks SVHC status for all additives.

"We lost a €2M contract because our supplier's EBS contained a trace impurity not on the old SDS. We didn't know it was now an SVHC. Don't rely on last year's data."

Compliance Officer, Electronics Manufacturer, Germany

12.3 Volatility and Emissions

Some dispersants volatilize during extrusion or injection molding at elevated temperatures. As a result, the production of the fume, odor, or particulate emissions take place that violates the workplace safety and environmental standards (e.g., OSHA, VOC limits, EU Industrial Emissions Directive). Fumes may be generated from PVC and engineering plastics and other materials that are extruded or molded at elevated temperatures

- High-Risk Additives:
  - Low-MW waxes (e.g., paraffin, some PE waxes) → evaporate above 200°C, causing smoke and respiratory hazards.
  - Fatty acid derivatives (EBS, stearates) → release amine or fatty acid vapors at >230°C, causing alarm systems in closed facilities to be triggered.
  - Surfactant-based systems → under long exposure to heat can decompose into formaldehyde or other VOCs.
- Consequences:
  - OSHA violations due to worker exposure.
  - Local environmental agency fines for VOC emissions.
  - Production halts during audits.
- Mitigation Strategies:
  - Use high-MW, thermally stable dispersants (e.g., Montan wax, PE-g-MA, Struktol® TPW) in high-temp applications (PA, PC, PBT).
  - Install effective venting and fume extraction systems, especially for PVC and engineering plastics.
  - Conduct fume analysis during process validation: Use GC-MS or portable VOC meters to identify emissions from new formulations.

"We thought our 220°C extrusion was fine, until the plant's air quality monitor flagged VOC spikes. Turns out, our 'universal' wax was boiling off. Switching to Montan wax cut emissions by 90%."

Process Safety Engineer, PVC Pipe Plant, Spain

12.4 Cross-Regulatory Insight: The Compliance Mindset

Compliance isn't just about checking boxes; it's about avoiding real business risk. This section cuts through the confusion with a practical mindset shift: if your additive works in the lab but fails on paper (or in audit), it fails everywhere. You'll learn to spot hidden regulatory traps, verify supplier claims, and build formulations that pass not only technical specs but global compliance requirements, before they cost you a shipment, a contract, or customer trust.

Table 12.4: Regulatory Misconceptions Vs Realities

Misconception	Reality
"If it works in the lab, it's fine."	Regulatory compliance is not about performance; it's about legal liability.
"My supplier said it's compliant."	Always verify with current, application-specific documentation (LOG, TDS, SVHC list). Generic claims are dangerous.
"We're not in Europe, so REACH doesn't matter."	Global supply chains mean your product may enter EU markets. Non-compliance blocks access.
"We can fix it later."	Once a batch is produced, you cannot "retrofit" compliance. You can only scrap, recall, or risk litigation.

12.5 Actionable Checklist for Regulatory Compliance

Compliance isn't something you check after the batch is made, it's built into every formulation decision. This checklist gives you a step-by-step verification process to ensure your dispersing and wetting additives meet food contact, emissions, and global chemical regulations before production starts. It's not about avoiding audits; it's about passing them with confidence, every time. Use this as your final go/no-go gate before scaling any new compound.

Before launching any new formulation, complete this checklist:

✅ Food Contact? → Is the additive listed in FDA 178.3820 or EU 10/2011? Do you have a valid LOG?

✅ Electronics/Industrial Use? → Is the additive REACH-compliant? Is it free of SVHCs above 0.1% w/w?

✅ High-Temp Processing? → Does the additive volatilize below your process temperature? Have you tested emissions?

✅ Global Market? → Are you compliant with both U.S. and EU regulations? What about RoHS, China RoHS, or California Prop 65?

✅ Documentation? → Is your technical file updated with supplier certificates, test reports, and change logs?

- Regulatory compliance is not a cost center, it is a competitive advantage.
- Formulations that meet global standards open doors to premium markets, reduce liability, and build customer trust.
- The right dispersant doesn't just improve dispersion, it ensures your product can be sold, used, and trusted around the world.

13. Example Commercial Products

Selecting the right dispersing and wetting additive is only half the battle. The other half is choosing a reliable, consistently available product from a supplier who understands your process, not just your chemistry. Below are industry-proven commercial products, each validated across multiple applications and backed by technical data sheets, regulatory documentation, and real-world performance. This is not a catalog. It's a shortlist of tools trusted by compounders on the front lines.

The commercial products are included to help readers contextualize the technical discussions in prior sections, such as chemistry, application, and performance, within real-world product formats. ***I am not affiliated with any manufacturer, distributor, or supplier of these products. No product is recommended, endorsed, or promoted herein.***

Table 13.1: Example Commercial Products (Trade Names, Suppliers, Chemistry)

Brand / Trade Name	**Supplier**	**Chemistry / Function**
Licowax PE 520	Clariant	Polyethylene wax – dispersant/lubricant
Hydrocerol®	Clariant	Wax-based dispersant for color masterbatches
Struktol® TPW Series	Schill + Seilacher	Multi-functional dispersing agents for thermoplastics
Atmer® 122	Croda	Antistatic + wetting agent for polyolefins
Silastic™ Additive MB Series	Dow	Silicone masterbatch dispersants
BYK-P 4100	BYK-Chemie	Wetting and dispersing additive for polar systems
EBS (Ethylene bis-stearamide)	Generic	Universal dispersant and external lubricant

This is not a "pick one and go" list. It's a decision matrix.

✅ When to Use What

Table 13.2: "When to Use What": Product Selection Guide by Goal

YOUR GOAL	RECOMMENDED PRODUCT
LOW-COST, GENERAL-PURPOSE DISPERSIONIN HDPE BAGS OR CRATES	EBS (but only if you accept limitations)
HIGH FILLER LOADING (40–60% $CACO_3$) IN PIPE OR PROFILES	PE-g-MA (1.5–2.0%) + Licowax PE 520 (0.2–0.5%), coupling + controlled lubrication
BLACK MASTERBATCH FOR AUTOMOTIVE TRIM	BYK-P 4100 (1.0–1.5%), eliminates specks, reduces dosage, prevents plate-out
FOOD-CONTACT COLOR MASTERBATCH	Hydrocerol® (0.7–1.0%), FDA-compliant, no migration, no blooming
SILICA-FILLED TPE DOOR SEALS	Silastic™ MB-100 (0.8–1.0%) + Tyzor® TPT (0.2%), silences surface reactivity, stops die build-up
FLAME-RETARDANT ABS FOR TELECOM HOUSINGS	Struktol® TPW 100 (1.2–1.5%), stabilizes BFR, improves flow, reduces loading
WHITE PP MEDICAL SYRINGES	BYK-P 4100 (1.0–1.2%) + HALS (0.3%), isolates TiO_2, prevents yellowing after sterilization

Red Flags to Avoid

- Using EBS as your "go-to" for everything, it's the Swiss Army knife that fails at every specialized task.
- Assuming "polyolefin-compatible" means "universal", PE-g-MA works in HDPE but not in PET.

- Ignoring supplier documentation, if they can't provide an FDA LOG or REACH compliance statement, walk away.
- Buying generic "EBS" from unknown vendors, trace impurities (e.g., phthalates, heavy metals) can invalidate your entire product line.

Pro Tip: Always Test in Your System

Even the best product can behave differently in your extruder, with your filler grade, or under your cooling rate.

Always validate with:

- A filter pressure rise test over 2 hours
- A speck count on molded plaques (10x magnification)
- A Delta-E measurement after aging or sterilization

The product that looks perfect on paper may fail in your line. The one that surprises you with performance? That's your new standard.

Final Thought: It's Not About the Brand, It's About the Fit

- The most expensive additive isn't always the best.
- The cheapest isn't always the worst.
- The right additive is the one that solves your specific problem, survives your process, meets your regulations, and pays for itself in reduced scrap.

Use this table not as a shopping list, but as a technical reference to guide your next formulation decision. When in doubt, ask your supplier:

"Which of your products has been validated in a twin-screw extruder at 240°C with 50% $CaCO_3$ and a 3-minute residence time?"

If they hesitate, or can't answer, keep looking.

14. Performance Evaluations

How do you know your dispersion is good? Not by hope, by measurement. Visual inspection, MFI, filter pressure rise, Delta-E, mechanical testing, this chapter details the five essential KPIs that turn subjective observation into objective data.

14.1 Visual Inspection & Speck Count

This is the first line of defense, simple, fast, and immediate. But only if done right. A quick glance isn't enough. You need controlled lighting, magnification (10x), and a known good standard to compare against. Whether it's a black masterbatch or a white medical component, visible specks are the most common sign of dispersion failure. The goal isn't just to spot them, it's to quantify them: ≤2 per 10 cm^2 for high-gloss parts, >5 means rework. This test separates acceptable batches from rejects before they leave the lab.

The first line of defense, but only if done right.

Method:

Produce test plaques (e.g., 100mm × 100mm × 2mm) via injection molding or compression molding under controlled conditions. Examine under 10x magnification (handheld loupe or low-power microscope) in uniform, diffuse lighting.

What to Count:

Count discrete, dark or colored particles ≥ 50 μm that are visibly distinct from the matrix. Exclude surface blemishes, flow lines, or mold marks.

Acceptance Criteria:

- Acceptable: ≤ 2 specks per 10 cm^2 (for high-gloss or transparent parts)

- Unacceptable: > 5 specks per 10 cm^2 (indicates incomplete wetting or poor stabilization)

Why It Matters:

Specks are the most visible symptom of dispersion failure. In automotive trim or medical devices, even one speck can trigger a batch rejection.

Pro Tip:

Always compare against a reference standard, a known good batch. Human eyes vary; standards don't.

14.2 Color Strength & Delta-E Measurement

Color consistency isn't just about matching a swatch, it's about process control and product integrity. This section shows how to use ΔE as a hard metric, not a guess. A high ΔE means poor dispersion, pigment agglomeration, or additive migration, especially in whites and pastels where even slight shifts trigger customer rejection. You'll learn how to measure color strength properly, interpret the numbers, and catch problems before they leave the lab. Because in real-world production, "close enough" only counts when the spectrophotometer agrees.

For pigmented systems, color is the most sensitive indicator of dispersion quality.

Method:

Use a spectrophotometer (e.g., BYK-micro, X-Rite) to measure color on molded plaques under D65 daylight illumination. Record L*, a*, b* values. Calculate ΔE*ab (total color difference) between the target and the sample:

$$\Delta Eab = \sqrt{[(L_2 - L_1)^2 + (a_2 - a_1)^2 + (b_2 - b^*_1)^2]}$$

Acceptance Criteria:

- Acceptable: ΔE ≤ 1.5 (industry standard for high-end applications)

- Unacceptable: ΔE > 2.0 (visible to the naked eye, especially in whites or pastels)

Critical Insight:

A low ΔE doesn't mean dispersion is good, it means color is consistent. A high ΔE always means something is wrong:

- Underdispersed pigment → low tinting strength → higher ΔE
- Agglomerates → light scattering → higher ΔE
- Migration → color shift over time → rising ΔE

Pro Tip:

Test ΔE after aging (e.g., 7 days at 60°C or after sterilization) to catch delayed migration or degradation.

14.3 Melt Flow Index (MFI) Changes

Melt Flow Index isn't just a number; it's a window into your dispersion quality. When additives do their job, they reduce interparticle friction and lower melt viscosity, which shows up as a higher MFI. But if you see a drop? That's a red flag: agglomerates are acting like fillers, increasing resistance and putting stress on your extruder. This test gives you a fast, repeatable way to spot dispersion issues, or confirm success, right on the production floor. Use it to catch problems early, optimize flow, and prove that your formulation is working as intended.

A simple, fast indicator of how dispersion affects melt rheology.

Method:

Measure MFI (ASTM D1238) at standard conditions (e.g., 230°C / 2.16 kg for PP). Compare the MFI of the compounded material to the base resin.

Interpretation:

- Increase in MFI → Improved flow due to better particle wetting and reduced interparticle friction (sign of good dispersion).
- Decrease in MFI → Indicates agglomeration (particles act as fillers, increasing viscosity) or degradation (chain scission from overheating).

Why It Matters:

In high-filler systems (e.g., 50% $CaCO_3$), a 10–15% increase in MFI after adding a dispersant is a strong signal that the additive is working.

Pro Tip:

Always test MFI at the same processing temperature as your production line. A drop in MFI during extrusion may signal re-agglomeration under shear.

14.4 Mechanical Testing (tensile, impact) to assess dispersion quality

Dispersion isn't just about how a part looks; it's about how it performs under stress. This section shows how tensile and impact testing reveal the real-world consequences of poor dispersion: weak spots, low elongation, and brittle failure. You'll learn how to use ASTM D638 and D256 not as pass/fail checks, but as diagnostic tools to confirm interfacial bonding and filler-polymer integration, because if your compound can't take the strain, it doesn't matter how good it mixes.

The ultimate test: Does dispersion affect the part's performance?

Method:

Run standard mechanical tests (ASTM D638 for tensile, ASTM D256 for impact) on injection-molded bars. Compare results to the base polymer and to a known good formulation.

What to Look For:

- Tensile Strength & Modulus: Improved with proper filler-polymer bonding (e.g., PE-g-MA in $CaCO_3$/HDPE).
- Impact Strength: Poor dispersion creates stress concentrators → lower impact values. A 10–20% drop often signals agglomeration.

Elongation at Break: Declines sharply when particles are poorly wetted and debond under load.

Critical Insight:

In engineering plastics, mechanical properties are the only metric that matters. A masterbatch with perfect color but low impact strength is a failure.

Pro Tip:

Use notched impact testing, it's far more sensitive to micro-defects than unnotched tests.

14.5 Filter Pressure Rise Test during extrusion

This test doesn't just monitor filtration, it predicts downtime. In real-world extrusion, a rising pressure drop across the screen pack is an early warning sign of agglomeration and poor dispersion. A stable filter pressure means your additive system is working. A rapid increase? That's clogged screens, unplanned shutdowns, and lost output. This section shows how to use filter pressure rise as a live diagnostic tool, because in high-speed pipe, film, or masterbatch lines, preventing screen changes is not about luck. It's about control.

The most underused, but most telling, test for industrial processors.

Method:

Install a standard mesh screen (e.g., 100–200 mesh) in the extruder's screen pack. Record the pressure drop across the screen (ΔP) every 30 minutes during a continuous 2-hour run.

Interpretation:

- Low, stable ΔP → Good dispersion, minimal agglomerates.
- Rapid ΔP increase (>10–15 bar over 2 hours) → Agglomerates are clogging the screen.

Why It Matters:

This test directly correlates to production downtime. A screen change every 4 hours? You're losing money. A stable ΔP for 24+ hours? Your dispersion is working.

Pro Tip:

Conduct this test before and after a formulation change. A 50% reduction in ΔP rise = a 50% reduction in maintenance cost.

14.6 Putting It All Together: The Evaluation Protocol

You don't manage what you don't measure, and you don't scale what isn't repeatable. This section brings together all five key tests, speck count, ΔE, MFI, mechanical performance, and filter pressure rise, into a single, field-ready validation protocol. It's the checklist top compounders use before releasing a new formulation: not to prove it looks good, but to prove it performs consistently, batch after batch. If you're serious about quality, this is your go/no-go gate.

For any new formulation or process change, implement this minimum validation protocol:

Table 14.6: Performance Evaluation Protocol: Tests & Timing

Step	Test	When To Run
1	Visual Speck Count	After initial molding, quick screen

2	Color Strength (ΔE)	After aging (24–72 hrs), for color-critical parts
3	MFI	During process validation, to confirm flow behavior
4	Filter Pressure Rise	During extrusion trial, to predict downtime
5	Mechanical Testing	Final validation, before production release

Rule of Thumb:

If you can't measure it, you can't control it. If you can't control it, you can't scale it.

These five evaluation methods are not academic exercises; they are production tools used daily by top-tier compounders. They turn guesswork into data, and data into decisions.

15. Example Formulations

Theory meets practice here. We provide ten ready-to-use formulations for pigment masterbatches, filled PP, PLA, PVC, PBT, and more, complete with dosages, processing tips, KPIs, and safety notes. These are templates built for real production environments.

15.1 Dispersing Agent for Pigment Masterbatch (PE-Based)

This formulation is engineered for high-loading, high-performance black or colored masterbatches in low-density polyethylene (LDPE) or linear low-density polyethylene (LLDPE) systems. Its purpose is to deliver complete pigment wetting, exceptional color strength, and zero specking, even under high-speed extrusion or injection molding conditions. Unlike generic blends, this system balances dispersion, lubrication, and thermal stability to meet the demands of packaging, film, and consumer goods applications where visual quality and process reliability are non-negotiable.

"This isn't just a color concentrate; it's a precision tool for consistent surface finish and throughput."

Masterbatch Formulator, Packaging Manufacturer, USA

15.1.1 Composition and Functionality

This formulation delivers a high-performance, cost-effective dispersing system tailored for PE-based pigment masterbatches, engineered not just to mix, but to perform under real-world extrusion and molding conditions. Built around a core of low-molecular-weight polyethylene wax and calcium stearate, it ensures rapid pigment wetting, minimal specking, and stable flow without the migration issues common with EBS or amide-based additives. Designed for compatibility with LDPE/LLDPE carriers and indirect food-contact compliance, it's a practical, no-compromise solution for packaging, film, and injection-molded applications where visual quality and process reliability are non-negotiable.

Table 15.1.1: PE-Based Pigment Masterbatch: Composition & Functionality

Ingredient	CAS #	Percentage (%)	Function
Polyethylene Wax (Low MW)	9002-88-4	3.0	Dispersant / Lubricant
Calcium Stearate	1592-23-0	1.0	Wetting Agent / Lubricant
Pigment (TiO_2 or Organic)	13463-67-7	50.0	Colorant
LDPE Carrier Resin	9002-88-4	46.0	Base Resin

Why This Blend Works:

- The PE wax provides immediate wetting and shear reduction.
- Calcium stearate enhances pigment surface coverage without migrating excessively.
- The LDPE carrier ensures compatibility and prevents bloom during storage.
- No EBS, deliberately excluded to avoid migration-related blooming or plate-out in thin films.

15.1.2 Processing Recommendations

For consistent, high-quality output in PE-based pigment masterbatches, processing isn't just about temperature, it's about timing, sequence, and shear control. This section lays out the practical extrusion parameters that seasoned compounders rely on: twin-screw profiles that balance melt stability with pigment wetting, residence times long enough to deagglomerate stubborn colorants but short enough to avoid degradation, and feeding strategies that prevent bridging and uneven distribution. No theory, just field-tested settings that keep screens clean, specks low, and your line running through the graveyard shift without surprises.

- Extruder Type: Twin-screw extruder (L/D $\geq$ 40), co-rotating.
- Temperature Profile:
  - Feed Zone: 140–150°C
  - Melting Zone: 170–180°C
  - Metering Zone: 180–190°C

Do not exceed 195°C, organic pigments and PE wax degrade above this threshold.

- Screw Speed: 80–100 RPM, sufficient for deagglomeration without fiber shear (if using fine pigments).
- Residence Time: Minimum 2.5 minutes, critical for full pigment wetting.
- Venting: Apply vacuum at Zone 3 if using moisture-sensitive pigments (e.g., carbon black).
- Pelletizing: Use water-cooled strand pelletizer. Cool pellets rapidly to lock in dispersion.
- Dosing: Feed pigment and dispersants via side feeder or pre-blend dry. Do not feed dry pigment through the main hopper, causes bridging and uneven distribution.

15.1.3 Key Performance Indicators (KPIs)

This section cuts through the noise with the five KPIs that actually matter on the plant floor: speck count under magnification, Delta-E after aging, MFI stability, filter pressure rise over a 2-hour run, and batch-to-batch repeatability. These aren't lab curiosities; they're the same metrics top-tier masterbatch producers use to avoid rework, pass automotive audits, and keep extrusion lines running 24/7 without surprises. If it doesn't show up in these numbers, it doesn't count.

Table 15.1.3: PE Masterbatch: Key Performance Indicators (KPIs)

KPI	Target	Measurement Method
Speck Count	≤ 2 specks per 10 cm^2 (10x magnification)	Visual inspection on injection-molded plaques
Color Strength (ΔE^*ab)	≤ 1.2 (vs. standard)	Spectrophotometer (D65 illuminant)
Melt Flow Index (MFI)	4.0–5.5 g/10min (230°C, 2.16 kg)	ASTM D1238
Filter Pressure Rise	≤ 5 bar over 2 hours	Extruder screen pack monitoring
Storage Stability (7 days @ 60°C)	$\Delta E \leq 0.8$	Accelerated aging test
Dispersion Consistency (Batch-to-Batch)	±0.3 ΔE	Statistical process control (SPC)

A ΔE > 1.5 or filter pressure > 10 bar over 2h = formulation failure.

15.1.4 Compatibility & Synergy

Formulation isn't just about what you put in, it's about how everything works together on the line. This section cuts through the chemistry to show which additives actually cooperate (like calcium stearate and PE wax in LDPE masterbatches) and which combinations quietly cause trouble (like EBS with silicone or zinc stearate in PVC). You'll get a no-nonsense map of what pairs well, what to avoid, and why, so your screen packs stay clean, your color stays consistent, and your compound doesn't spend its life in rework.

✅ Synergistic With:

- Titanium Dioxide (TiO_2), PE wax and CaSt synergize to fully wet high-surface-area rutile TiO_2.
- Organic Pigments (e.g., Pigment Red 177), Low-MW wax prevents agglomeration without blooming.
- Antioxidants (e.g., Irganox 1010), No interference; stable under processing conditions.

✕ Avoid Combining With:

- EBS or Erucamide, Causes bloom and migration, especially in thin films or food-contact applications.
- Silicone Additives, Can cause surface defects (fish-eyes) in blown film.
- Polar Carriers (e.g., EVA, PA), Incompatible; will phase-separate.

Pro Tip: This blend is optimized for non-polar carriers only. For PET or PA masterbatches, switch to BYK-P 4100 + PEG ester.

15.1.5 Advantages / Highlights

This formulation delivers consistent, industrial-grade performance without over-engineering or unnecessary cost. It eliminates visible specks even at 50% pigment loading, critical for thin-wall packaging and high-gloss applications where surface defects trigger batch rejection. The blend avoids EBS and other low-molecular-weight migrators, so there's no plate-out during extended extrusion runs and no bloom in storage. Melt flow remains stable, keeping processing predictable on standard LDPE/LLDPE lines, and all components are listed under FDA 21 CFR 178.3820 for indirect food-contact compliance. In short: it's a lean, reliable system that does exactly what it needs to, no more, no less.

- Zero specking even at 50% pigment loading, critical for transparent or white films.

- Low MFI increase, maintains processability without sacrificing mechanical properties.
- No plate-out, calcium stearate is non-migrating at this dosage and temperature.
- Cost-effective, avoids expensive specialty dispersants while delivering industrial-grade performance.
- FDA-compliant, all components listed under 21 CFR 178.3820 for indirect food contact.

15.1.6 Typical Applications

This PE-based pigment masterbatch formulation example is built for real-world use where consistent color, clean processing, and surface quality matter more than lab-perfect specs. It's commonly used in LDPE/LLDPE films, like stretch wrap, grocery bags, and food packaging, where even a few specks can trigger customer rejection. It also performs reliably in injection-molded containers (food storage, cosmetic tubs) and thermoformed trays that demand smooth, high-gloss finishes without plate-out or blooming. Because it avoids EBS and other migratory additives, it's suitable for indirect food-contact applications and thin-gauge films that would otherwise suffer from surface defects. In short: if your part needs to look clean, run cleanly, and stay clean through storage and conversion, this is the baseline to start from.

- Stretch Film (LDPE/LLDPE), for retail, food packaging
- Trash Bags & Liners, black or colored, high-gloss finish
- Injection-Molded Containers, food storage, cosmetic packaging
- Thermoformed Trays, where surface clarity and color consistency are critical

Ideal for applications where visual defects = rejected batches.

15.1.7 Handling & Safety Notes

Handle this formulation like any industrial compound: respect the dust, control the heat, and protect the people. Pigments, especially TiO_2, should be treated as inhalation hazards; use N95 masks and local exhaust ventilation during dry blending. Calcium stearate and polyethylene wax are low-toxicity, but fine powders can irritate eyes and lungs, so wear safety glasses and gloves. Keep the blend away from open flames, LDPE and waxes are combustible. Store in a dry, cool place to avoid moisture uptake, which leads to specks and poor dispersion downstream. No exotic risks here, but don't get complacent: consistent safety isn't about drama, it's about discipline on the floor.

- Dust Control: Pigment handling requires N95 masks and local exhaust ventilation (especially for TiO_2, IARC Group 2B).
- Skin Contact: Calcium stearate and PE wax are non-irritating but may cause mechanical irritation if inhaled as dust.
- Fire Hazard: LDPE and waxes are combustible. Store away from open flames.
- PPE: Gloves, safety glasses, and dust mask required during dry blending.
- No known acute toxicity, SDS compliant with GHS standards.

15.1.8 Packaging and Storage

Package form is typically 25 kg multi-wall paper bags with a polyethylene liner. Avoid storing more than three bags high, excessive stacking can rupture seams and expose material to ambient humidity. Shelf life is 12 months under proper conditions, but always inspect for caking or discoloration before use. If the pellets feel dusty or show signs of surface bloom, test a small batch before full-scale processing.

Store this masterbatch in a cool, dry place, ideally between 15–25°C and below 50% relative humidity. Moisture is the enemy: it causes pigment clumping, which leads to specks and inconsistent color downstream. Keep bags sealed until use, and follow first-in, first-out (FIFO) inventory practices to avoid aging-related performance drift.

- Packaging: 25 kg multi-wall paper bags with PE liner.
- Storage Conditions:
  - Temperature: 15–25°C
  - Humidity: <50% RH, moisture causes pigment clumping and poor dispersion.
- Shelf Life: 12 months if stored dry and sealed.
- Handling: Use first-in, first-out (FIFO) inventory. Avoid stacking more than 3 bags high to prevent bag rupture.

15.1.9 Optimization Tips / Remarks

This isn't a set-it-and-forget-it formulation, it's a starting point that rewards attention to detail. Small tweaks in pigment drying, screw speed, or additive balance can mean the difference between batch rejection and consistent performance, especially when pushing pigment loadings above 50% or running thin-gauge film. The tips here reflect what actually works on the line: how to troubleshoot specks that only show up after two weeks of storage, when to add Montan wax as a flow aid without risking bloom, and why upgrading to a specialty dispersant like BYK-P 4100 pays off in high-value applications. Treat this as your field notes, not a manual, but a conversation with someone who's already run the trials and made the mistakes so you don't have to.

Optimization Tips

- For Higher Pigment Loadings (60–70%): Increase PE wax to 4.0% and reduce LDPE carrier to 35%. Monitor MFI, if it drops below 3.0, add 0.2% Montan wax for flow aid.
- For Transparent Films: Use rutile TiO_2 (not anatase), lower refractive index, less haze.

- For Food Contact: Verify that your pigment supplier provides a Letter of Guarantee (LOG) for FDA compliance, even if the dispersant is compliant, the pigment may not be.
- Troubleshooting Specks:
  - If specks appear after 2 weeks of production → Check for moisture in pigment. Dry at 80°C for 4 hours.
  - If specks appear only in thin-gauge film → Increase residence time by reducing screw speed to 70 RPM.
- Upgrade Path: For premium automotive or medical applications, replace PE wax + CaSt with BYK-P 4100 (1.5%), eliminates 90% of variability and enables thinner films.

Remarks

This formulation is not a starting point, it's a battle-tested baseline.
Use it as a template. Test it. Optimize it. But never assume it's "good enough" without verifying KPIs on your own line.

15.2 Dispersing and Wetting Additive for Talc-Filled PP

This formulation is engineered for high-performance, talc-filled polypropylene (PP) compounds used in automotive, appliance, and industrial applications where stiffness, dimensional stability, and surface finish are critical. Talc ($Mg_3Si_4O_{10}(OH)_2$) is a high-aspect-ratio filler that improves rigidity and heat distortion temperature, but it is notoriously difficult to wet and disperse uniformly in non-polar PP matrices. Without proper coupling, talc agglomerates, increases melt viscosity, reduces impact strength, and causes surface defects.

This formulation uses PP-g-MA as a coupling agent to chemically bond talc to the PP matrix, while zinc stearate provides just enough lubrication to ensure smooth flow without compromising dispersion. The result is a balanced compound that delivers high stiffness, low warpage, and consistent surface quality, even at 40% loading.

"Talc isn't just a filler; it's a structural element. If it's not bonded, it's a defect waiting to happen."

Materials Engineer, Automotive Tier-1, Germany

15.2.1 Composition and Functionality

This talc-filled PP formulation isn't just about loading in mineral; it's about making every particle work. Built around 3.0% PP-g-MA (maleated polypropylene), it chemically bonds talc to the polymer matrix instead of just pushing it through the extruder. A light touch of zinc stearate (0.5%) keeps flow smooth without softening the compound or hurting heat resistance. No EBS, no waxes, those might help flow, but they sabotage the interface. The result is a stiff, dimensionally stable PP that runs cleanly, holds tight tolerances, and won't warp or chalk on the customer's line. If your talc is just filler, you're wasting performance. Here, it's reinforcement.

Table 15.2.1: Talc-Filled PP: Composition & Functionality

Ingredient	CAS #	Percentage (%)	Function
PP-g-MA (Maleic Anhydride Grafted Polypropylene)	N/A	3.0	Coupling Agent / Dispersant
Zinc Stearate	557-05-1	0.5	Wetting / Flow Aid
Talc	14807-96-6	40.0	Functional Filler
Polypropylene Homopolymer	9003-07-0	56.5	Base Polymer

Why This Blend Works:

- PP-g-MA is the only dispersant that chemically bonds to talc in PP, unlike waxes or EBS, which merely coat particles.
- Zinc stearate is used sparingly because higher levels can soften the matrix and reduce HDT.

- The low MFI PP ensures the compound doesn't become too fluid during processing, preserving filler orientation and stiffness.

15.2.2 Processing Recommendations

Getting talc to stay put in PP isn't about brute force, it's about timing, temperature, and chemistry working together. This section gives you the exact extrusion profile that keeps 40% talc loading from agglomerating, clogging screens, or weakening the part: a cool feed zone to avoid premature melting, a precise 3-minute residence time for PP-g-MA to bond to talc surfaces, and just enough zinc stearate to lubricate without softening the matrix. No guesswork, just the settings that deliver stiff, dimensionally stable parts with clean surfaces and consistent mechanicals, batch after batch. If you're still running this on a single-screw extruder, you're wasting talc.

- Extruder Type: Twin-screw co-rotating extruder ($L/D \geq 40$), with high shear mixing zones.
- Temperature Profile:
  - Feed Zone: 170–180°C (keep cool to avoid premature melting of talc agglomerates)
  - Melting Zone: 200–210°C (ensure PP-g-MA activates without degradation)
  - Metering Zone: 205–215°C (avoid exceeding 220°C, PP-g-MA degrades above this)
- Screw Speed: 100–130 RPM, sufficient to break talc clusters without excessive fiber breakage.
- Residence Time: Minimum 3 minutes, critical for PP-g-MA to fully react with talc surface.
- Venting: Apply vacuum at Zone 3 if talc contains residual moisture (>0.05%).

- Pelletizing: Use underwater pelletizer with chilled water (15–20°C) to rapidly solidify melt and lock in dispersion.
- Dosing: Pre-blend talc and PP-g-MA before feeding into extruder. Feed zinc stearate via side feeder at Zone 2 to avoid early migration.

Critical Note: Do not use single-screw extruders for this formulation, insufficient shear will leave talc agglomerates intact.

15.2.3 Key Performance Indicators (KPIs)

You don't know if your talc-filled PP is performing unless you measure what matters on the line. This section cuts through subjective claims with the five KPIs that actually predict success: tensile modulus (≥2,200 MPa), heat deflection temperature (≥145°C), notched impact strength (≥8 kJ/m²), warpage (≤±0.3 mm over 100 mm), and filter pressure rise (≤6 bar over 2 hours). These aren't lab curiosities; they're the same metrics automotive and appliance makers use to accept or reject a batch. If your compound misses one, it's not "close enough", it's scrap. Measure it, control it, ship it.

Table 15.2.3: Talc-Filled PP KPIs

KPI	Target	Measurement Method
Tensile Modulus (ASTM D638)	≥ 2,200 MPa	Mechanical testing of injection-molded bars
Heat Deflection Temp (HDT @ 1.82 MPa)	≥ 145°C	ASTM D648
Impact Strength (Notched, 23°C)	≥ 8 kJ/m²	ASTM D256
Warpage (Flatness)	≤ ±0.3 mm over 100mm	Laser displacement sensor
Visual Surface Quality	No streaks, specks, or roughness	Visual inspection under 10x magnification

Filter Pressure Rise	≤ 6 bar over 2 hours	Extruder screen pack monitoring
Batch-to-Batch Consistency	±5% variation in modulus	SPC control chart

A drop in impact strength below 7 kJ/m² or HDT below 140°C = poor interfacial bonding. Revisit coupling agent dosage or drying.

15.2.4 Compatibility & Synergy

In talc-filled PP, performance lives or dies by what you mix, and what you don't. This section cuts through trial-and-error with a field-tested map of what works: how PP-g-MA bonds chemically with talc while zinc stearate lubricates just enough without softening the matrix, and why adding EBS or calcium stearate, even with good intentions, undermines coupling and hurts stiffness. You'll also see which additives team up reliably (like PP-g-MA with glass fiber or nucleating agents) and which combinations quietly sabotage surface quality or paint adhesion (silicones, anyone?). No theory, just the practical pairings that keep your compound stiff, clean, and stable from extrusion to end use.

✅ Synergistic With:

- Talc (fine-grain), PP-g-MA chemically bonds to surface hydroxyls, maximizing reinforcement.
- Glass fibers (up to 10%), PP-g-MA also improves fiber-matrix adhesion.
- Antioxidants (Irganox 1010), No interference; stable under processing conditions.
- Nucleating agents (e.g., Sorbitol-based), Enhance crystallinity and HDT without affecting dispersion.

❌ Avoid Combining With:

- EBS or PE Wax, Competes with PP-g-MA for surface sites; reduces coupling efficiency.
- Calcium Stearate, Can interfere with PP-g-MA reaction; may increase viscosity unpredictably.
- Silicone-based additives, May migrate to surface and reduce paintability or weld strength.
- High-MFI PP (MFI > 25), Too fluid; reduces filler orientation and stiffness.

Pro Tip: Never add PP-g-MA after the filler. Always pre-blend or feed simultaneously, delayed addition results in incomplete reaction.

15.2.5 Advantages / Highlights

This talc-filled PP formulation delivers the stiffness, thermal resistance, and dimensional stability demanded in automotive and industrial parts, without the headaches of poor surface finish or batch inconsistency. The secret isn't just the 40% talc loading; it's the 3.0% PP-g-MA that chemically bonds talc to the matrix, eliminating weak interfaces that cause warpage, poor paint adhesion, or filler pull-out. Zinc stearate is kept to a lean 0.5%, just enough to aid flow but not so much that it softens the compound or lowers heat deflection temperature. The result? Clean extrusion, minimal screen pressure rise, and parts that stay flat and strong, even after thermal cycling. No plate-out, no bloom, no guesswork, just repeatable performance where it counts.

- Highest stiffness-to-weight ratio among mineral-filled PP systems.
- Low warpage due to uniform talc orientation and strong interfacial bonding.
- Improved paintability and adhesion, no blooming or surface contamination.
- No plate-out, zinc stearate is used at minimal, non-migrating dosage.
- Cost-effective, avoids expensive engineering resins while achieving near-PA6 performance.

- Recyclable, all components are compatible with PP recycling streams.

15.2.6 Typical Applications

This talc-filled PP compound is built for parts that need to stay stiff, flat, and stable under heat and load, without the cost or complexity of glass fiber or engineering resins. It's commonly used in automotive interiors (dashboards, door panels), appliance housings (washer drums, dryer liners), and industrial components (pallets, bins) where warpage, surface finish, and thermal resistance directly impact assembly and performance. Because the formulation uses PP-g-MA instead of generic lubricants, it maintains paint adhesion, withstands thermal cycling, and avoids the plate-out that plagues EBS-based systems. If your part has tight dimensional tolerances, needs to hold its shape after molding, or must pass automotive audits on first run, this is the practical starting point.

- Automotive: Dashboard substrates, door panels, instrument clusters, under-hood components
- Appliances: Washing machine drums, dryer housings, refrigerator liners
- Industrial: Pallets, bins, tool handles, electrical enclosures
- Consumer Goods: Storage containers, furniture parts, outdoor equipment

Ideal for parts requiring high rigidity, thermal resistance, and dimensional accuracy, without the cost of glass fiber or engineering resins.

15.2.7 Handling & Safety Notes

Handle this talc-filled PP compound like any industrial material: respect the dust, control the exposure, and protect your team. Talc can generate fine respirable particles, wear N95 masks and use local exhaust ventilation during blending or feeding. Zinc stearate and PP-g-MA are low-hazard, but airborne powders can still irritate eyes and lungs, so safety glasses and gloves are non-negotiable. Store away from ignition sources, PP is combustible, and talc dust can be explosive in high concentrations. Keep bags sealed and dry; talc absorbs moisture, which leads to clumping, poor dispersion, and steam-induced voids during processing. No exotic risks here, but skip the basics, and you'll pay in downtime, defects, or worse.

- Dust Control: Talc is a fine powder, use N95 masks and local exhaust ventilation during handling (IARC Group 2B for respirable fibers).
- Skin Contact: Zinc stearate and PP-g-MA are non-irritating but may cause mechanical irritation if inhaled as dust.
- Fire Hazard: PP is combustible. Store away from open flames.
- PPE Required: Gloves, safety glasses, and dust mask during dry blending.
- No known acute toxicity, SDS compliant with GHS standards.

15.2.8 Packaging and Storage

Packaging is typically 25 kg multi-wall paper bags with a polyethylene liner for talc and PP-g-MA, and 20 kg HDPE drums for zinc stearate. Avoid stacking more than three bags high to prevent seam rupture. Shelf life is 12 months for PP-g-MA (if kept dry), indefinite for talc, and 24 months for zinc stearate. Before use, inspect for caking, off-color, or loss of free-flowing powder, any of these signs mean the material needs testing before full-scale processing.

Store this talc-filled PP compound in a dry, climate-controlled space, ideally 15–25°C and below 40% relative humidity. Talc absorbs moisture readily, and PP-g-MA is sensitive to hydrolysis; both will cause clumping, poor dispersion, and pressure spikes during extrusion if exposed to humidity. Keep bags sealed until use and follow strict FIFO (first-in, first-out) inventory practices, PP-g-MA loses reactivity over time, especially if moisture gets in.

- Packaging:
  - PP-g-MA and talc: 25 kg multi-wall paper bags with PE liner.
  - Zinc stearate: 20 kg HDPE drums.

- Storage Conditions:
  - Temperature: 15–25°C
  - Humidity: <40% RH, moisture reduces PP-g-MA reactivity and causes talc clumping.
- Shelf Life:
  - PP-g-MA: 12 months (degrades with moisture exposure)
  - Talc: Indefinite if kept dry
  - Zinc Stearate: 24 months
- Handling: Use FIFO inventory. Store PP-g-MA in sealed, desiccated containers.

15.2.9 Optimization Tips / Remarks

This talc-filled PP system works reliably as-is, but it rewards smart tweaks. If warpage creeps in at high cycle speeds, it's often not the talc but uneven cooling or screw design; adjust mold temperature or add reverse kneading elements before touching the formulation. If impact strength drops below 8 kJ/m², consider blending in 1–2% EPR rather than cutting talc, stiffness stays, toughness improves. For painted parts, skip silicone additives entirely, even in trace amounts; they'll pass visual inspection but fail adhesion tests down the line. And always verify talc dryness: just 0.1% moisture can spike filter pressure and create micro-voids that show up only after thermal cycling. This isn't about fixing what's broken, it's about fine-tuning a solid baseline so it holds up under real-world stress, shift after shift.

Optimization Tips

- For Higher Talc Loadings (45–50%): Increase PP-g-MA to 3.5–4.0%. Monitor MFI, if it drops below 8 g/10min, add 0.2% PE wax as a *processing aid* (not a dispersant).

- For Improved Impact Strength: Add 1–2% ethylene-propylene rubber (EPR) as an impact modifier, PP-g-MA still couples talc, while EPR absorbs energy.
- Troubleshooting Warpage:
  - If warpage exceeds ±0.5 mm → Check talc particle size. Use <3 μm for better orientation.
  - If warpage is directional → Increase screw speed and use reverse kneading elements to randomize flow.
- For Painted Parts: Avoid silicone additives entirely, even 0.1% can cause paint adhesion failure.
- Upgrade Path: For high-temperature applications (>120°C), replace PP-g-MA with PP-g-MA + 0.5% titanate coupling agent (Tyzor® TPT) for enhanced thermal stability.
- Quality Control Tip: Run a DSC scan to confirm crystallization temperature shift, successful coupling increases crystallinity and shifts peak by 2–5°C.

Remarks

This formulation is not a generic blend, it's a precision-engineered system designed for performance-critical applications.
Use it as a benchmark. Test it rigorously. Optimize it only after validating KPIs on your own line.

15.3 Dispersant System for Color Masterbatch (PS-Based)

This formulation is engineered for high-clarity, high-gloss color masterbatches in polystyrene (PS), including GPPS (general-purpose polystyrene) and HIPS (high-impact polystyrene), used in applications such as disposable cutlery, cosmetic containers, appliance housings, and transparent packaging. PS is a rigid, amorphous polymer with low surface energy and limited polarity, making it challenging to wet and disperse pigments uniformly without compromising optical clarity.

Unlike polyolefins, PS cannot tolerate high levels of waxes or amide-based lubricants, they cause haze, bloom, or surface defects. This system leverages stearic acid and EBS in a precisely balanced ratio to deliver excellent pigment wetting, zero specking, and exceptional surface gloss, without clouding or migration.

"In PS, you can't hide a bad dispersion with gloss. If the pigment isn't perfectly wetted, the customer sees it, every time."
Masterbatch Formulator, Consumer Goods, China

15.3.1 Composition and Functionality

This PS-based color masterbatch system is built for clarity, gloss, and consistency, not just in the lab, but on high-speed lines where defects show up instantly. It uses stearic acid and a lean dose of EBS to wet pigments without clouding the melt, paired with a pure PS carrier that won't phase-separate or bloom. No waxes, no metal soaps, no silicones, those might work in polyolefins, but in PS they cause haze, streaks, or surface tack that kills premium aesthetics. Every ingredient here has one job: disperse the pigment cleanly, lock in gloss, and stay invisible in the final part. If your application demands see-through quality or high-gloss finish, like cosmetic jars or cutlery, this is the baseline that won't let you down.

Table 15.3.1: PS-Based Color Masterbatch: Composition & Functionality

Ingredient	CAS #	Percentage (%)	Function
Polystyrene	9003-53-6	94.0	Base Resin
Stearic Acid	57-11-4	1.5	Lubricant / Dispersant
EBS (Ethylene Bis-Stearamide)	110-30-5	1.0	Wetting Agent
Pigment Concentrate	Various	3.5	Colorant

Why This Blend Works:

- Stearic acid is ideal for PS because it has sufficient polarity to wet pigments but low enough molecular weight to remain soluble in the melt, unlike EBS, which can bloom if overdosed.
- EBS is used at low dosage (1.0%) to avoid migration, critical in transparent PS where even 0.2% blooming causes haze.
- The PS carrier ensures full compatibility, no phase separation, no bloom, no surface defects.
- No silicone, no wax, no metal soaps, all of which are incompatible with PS clarity requirements.

15.3.2 Processing Recommendations

Processing PS-based color masterbatches isn't about brute force, it's about finesse. This section gives you the exact twin-screw profile that keeps organic pigments and TiO_2 fully wetted without clouding the melt or degrading the resin: a cool feed zone (180–190°C) to avoid premature softening, a narrow 200–215°C melt window to prevent thermal yellowing, and just enough residence time (2–2.5 minutes) for stearic acid and EBS to do their job without blooming. No water quenching, PS cracks under thermal shock, so air-cool strands and dry pigments to <0.05% moisture upfront. If you're still running this on a single-screw extruder or feeding dry pigment through the hopper, you're guaranteeing specks and haze. This is precision compounding, not bulk mixing.

- Extruder Type: Twin-screw co-rotating extruder (L/D ≥ 30). Single-screw systems risk incomplete dispersion due to low shear.
- Temperature Profile:
  - Feed Zone: 180–190°C (keep cool to avoid premature melting of pigment concentrate)
  - Melting Zone: 200–210°C (ensure full melt of PS without degradation)

- Metering Zone: 205–215°C (do not exceed 220°C, PS degrades above this)

Critical: PS is sensitive to thermal history. Avoid prolonged residence time.

- Screw Speed: 90–110 RPM, sufficient for pigment deagglomeration without shear degradation.
- Residence Time: 2.0–2.5 minutes, longer than PP/PE due to PS's higher melt viscosity.
- Venting: Apply vacuum at Zone 2 if using moisture-sensitive pigments (e.g., phthalocyanines).
- Pelletizing: Use air-cooled strand pelletizer. Avoid water quenching, PS is brittle when cooled too rapidly.
- Dosing: Pre-blend pigment concentrate with EBS and stearic acid before feeding into extruder. Feed PS carrier via main hopper.

Pro Tip: Always pre-dry pigment concentrate to <0.05% moisture, PS is hygroscopic and prone to voids if wet.

15.3.3 Key Performance Indicators (KPIs)

In PS-based color masterbatches, performance isn't judged by how it mixes, it's proven by what you can measure on the line. This section focuses on the six KPIs that matter most: haze ≤1.5% for clarity-critical packaging, speck count ≤1 per 10 cm² under 10x magnification, Delta-E ≤1.0 for batch-to-batch color consistency, surface gloss ≥85 GU for premium appearance, MFI stability between 7.5–9.5 g/10min to ensure smooth processing, and post-aging haze <1.8% after 7 days at 60°C. These aren't optional targets, they're the line between acceptance and rejection in cosmetics, food packaging, and consumer goods where optics equal quality. If your formulation doesn't hit these numbers, it doesn't ship.

Table 15.3.3: Key Performance Indicators (KPIs)

KPI	Target	Measurement Method
Visual Clarity (Haze %)	≤ 1.5%	ASTM D1003 (1mm molded plaque)
Speck Count (10x magnification)	≤ 1 speck per 10 cm²	Visual inspection under diffused light
Color Strength (ΔE*ab)	≤ 1.0 (vs. standard)	Spectrophotometer (D65 illuminant)
Surface Gloss (60°)	≥ 85 GU	Glossmeter (ASTM D523)
Melt Flow Index (MFI)	7.5–9.5 g/10min (200°C, 5kg)	ASTM D1238
Storage Stability (7 days @ 60°C)	ΔE ≤ 0.5, Haze ≤ 1.8%	Accelerated aging test
Batch-to-Batch Consistency	±0.2 ΔE	SPC control chart

A haze > 2.0% or ΔE > 1.5 = formulation failure. PS does not forgive dispersion flaws.

15.3.4 Compatibility & Synergy

In PS-based color masterbatches, getting the right look starts with keeping the wrong additives out. This section lays out what actually works: how stearic acid and EBS at low levels wet pigments without clouding the melt, and why even small amounts of PE wax, calcium stearate, or silicone will ruin clarity with haze, bloom, or streaks. It also shows where synergy helps, like pairing EBS with high-purity, surface-treated TiO_2 for brighter whites, and where it backfires, like using “universal” dispersants that phase-separate in PS. No theory, just the practical combinations that keep your film glossy, your cutlery clean, and your customer from rejecting the batch.

☑ Synergistic With:

- Organic Pigments (e.g., Pigment Yellow 13, Red 112), Stearic acid and EBS provide excellent wetting without clouding.
- Titanium Dioxide (TiO_2), Use only high-grade, surface-treated TiO_2. Uncoated grades cause haze.
- Antioxidants (Irganox 1010), No interference; stable under processing.
- Nucleating Agents (e.g., Sorbitol-based), Enhance crystallinity in HIPS without affecting clarity.

✕ Avoid Combining With:

- Polyethylene Wax or Montan Wax, Cause severe haze and bloom in PS.
- Silicone-Based Additives, Cause fish-eyes, poor printability, and paint adhesion failure.
- Calcium Stearate or Zinc Stearate, Form insoluble soaps with PS, leading to white streaks.
- EBS > 1.5%, Causes surface blooming within days, even in opaque grades.

Pro Tip: Never use "universal" dispersants like BYK-P 4100 in PS, it's polar and incompatible. Stick to stearic acid/EBS.

15.3.5 Advantages / Highlights

This PS-based color masterbatch system delivers the clarity, gloss, and color consistency demanded in premium packaging and consumer goods, without the haze, bloom, or plate-out that plagues generic formulations. By using just 1.5% stearic acid and 1.0% EBS, it fully wets pigments while staying soluble in the PS melt, avoiding the surface defects caused by waxes, silicones, or metal soaps. The result is a clean, high-gloss finish that holds up under injection molding, stays stable during storage, and meets FDA requirements for indirect food contact. No over-engineering, no hidden trade-offs, just reliable performance where appearance is non-negotiable.

- Exceptional clarity, Haze levels meet premium packaging and cosmetic container standards.
- Zero bloom, even after 30 days at 60°C, surface remains clean.
- High gloss, achieves >85 GU without polishing or post-treatment.
- Low dosage, only 2.5% total additive load minimizes cost and process impact.
- Food-contact compliant, Stearic acid and EBS are FDA-listed for indirect food contact.
- Easy processing, Compatible with standard PS extrusion and injection molding lines.

15.3.6 Typical Applications

This PS-based color masterbatch system is built for parts where gloss, clarity, and color accuracy can't be compromised, like cosmetic jars, disposable cutlery, appliance housings, and transparent food packaging. It's formulated specifically for polystyrene's narrow processing window and low tolerance for haze, using only additives that stay dissolved in the melt and won't bloom or cloud the surface. If your application demands a clean, high-gloss finish straight off the mold, without post-treatment, polishing, or customer rejects, this is the baseline that delivers, batch after batch.

- Transparent Packaging: Cosmetic jars, food containers, blister packs
- Disposable Cutlery & Cups, High-gloss, colored PS for retail branding
- Appliance Housings, White or colored PS for coffee makers, toasters
- Toys & Consumer Goods, Bright, glossy parts requiring visual consistency
- Medical Devices, Clear PS components for diagnostic kits (non-sterile)

Ideal for applications where "see-through" aesthetics are part of the brand identity.

15.3.7 Handling & Safety Notes

Handle this PS-based color masterbatch like any industrial compound: respect the dust, protect the people, and control the basics. Organic pigments and stearic acid are low-hazard, but fine powders can irritate eyes and lungs, wear safety glasses, gloves, and an N95 mask during dry handling. PS is flammable, so keep material away from ignition sources and avoid open flames. Store in a dry, cool place; PS absorbs moisture, and even small amounts can cause voids or surface defects during processing. No exotic risks here, but skip the fundamentals, and you'll pay in downtime, scrap, or safety incidents. Good handling isn't about paperwork. It's about keeping your team and your line running without surprises.

- Dust Control: Pigment handling requires N95 masks and local exhaust ventilation (especially for organic pigments).
- Skin Contact: Stearic acid and EBS are non-irritating. Avoid prolonged skin contact with dry powders.
- Fire Hazard: PS is flammable. Store away from ignition sources.
- PPE Required: Gloves, safety glasses, and dust mask during dry blending.
- No known acute toxicity, SDS compliant with GHS standards.

15.3.8 Packaging and Storage

Store this PS-based color masterbatch in a cool, dry place, ideally 15–25°C and under 40% relative humidity. Polystyrene is hygroscopic, and moisture leads to voids, surface bubbles, and inconsistent pigment dispersion during processing. Keep bags sealed until use and follow strict FIFO (first-in, first-out) inventory practices, PS can yellow over time if exposed to heat or ambient UV.

Packaging is typically 25 kg multi-wall paper bags with a polyethylene liner for the masterbatch, and 10–20 kg sealed HDPE pails for pigment concentrate (which is especially moisture-sensitive). Avoid stacking more than three bags high to prevent seam rupture. Shelf life is 18 months for PS carrier and 12 months for pigment concentrate, provided it stays dry and dark.

Before use, inspect for clumping, discoloration, or loss of free-flowing powder. If in doubt, dry at 70°C for 2–3 hours and test a small batch. This isn't just raw material; it's a precision color tool. Treat it like one

- Packaging:
    - PS carrier and additives: 25 kg multi-wall paper bags with PE liner.
    - Pigment concentrate: 10–20 kg sealed HDPE pails (moisture-sensitive).
- Storage Conditions:
    - Temperature: 15–25°C
    - Humidity: <40% RH, PS absorbs moisture, causing voids and bubbles during processing.
- Shelf Life:
    - PS Carrier: 18 months
    - Pigment Concentrate: 12 months (if stored dry)
    - Stearic Acid / EBS: 24 months
- Handling: Use FIFO inventory. Store pigment concentrate in desiccated environment.

15.3.9 Optimization Tips / Remarks

This PS-based color masterbatch system works reliably as-is, but it rewards careful tuning. If haze creeps in during high-output runs, it's often not the pigment, but EBS migrating to the surface; try cutting EBS to 0.8% and relying more on stearic acid for wetting. For deep reds or blues with high-tinting organic pigments, pre-disperse the colorant with stearic acid before compounding, it prevents specking without adding cost. Always verify pigment dryness: PS is hygroscopic, and even 0.03% moisture causes micro-voids that scatter light and kill gloss. And skip silicone additives entirely if the part will be printed or painted, they pass visual inspection but fail adhesion tests down the line. This isn't about fixing what's broken, it's about tightening a solid baseline so it holds up under real-world processing, shift after shift.

Optimization Tips

- For Higher Pigment Loadings (5–7%): Increase stearic acid to 2.0% and EBS to 1.2%. Monitor MFI, do not let it drop below 6.0 g/10min.
- For Opaque Grades: You may replace stearic acid with 0.5% calcium stearate for better opacity, but test for bloom risk.
- Troubleshooting Haze:
    - If haze appears after 2 weeks → Check for EBS > 1.0%. Reduce to 0.8%.
    - If haze appears during molding → Increase barrel temperature by 5°C to improve pigment wetting.
- For UV Stability: Add 0.1–0.2% HALS (e.g., Tinuvin 770) if used outdoors, PS degrades under UV.
- Upgrade Path: For ultra-high-clarity applications (e.g., optical lenses), replace PS carrier with PMMA and use BYK-P 4100, but only if color is not critical.
- Quality Control Tip: Run a FTIR scan of the pellet surface, presence of free stearic acid indicates under-mixing or poor dispersion.

Remarks

This formulation is not a generic blend, it's a precision-engineered system designed for the unforgiving clarity requirements of polystyrene.

Use it as your baseline for high-end PS color applications.

Test rigorously. Optimize only after validating KPIs on your own line.

15.4 Wetting Agent for PVC Compounds

This formulation is engineered for rigid and semi-rigid PVC compounds used in window profiles, pipe fittings, electrical conduits, and medical tubing, applications demanding uniform pigment/filler dispersion, long-term thermal stability, and smooth surface finish. PVC is inherently challenging to process: it has low melt flow, high viscosity, and is highly sensitive to thermal degradation. Pigments and stabilizers often agglomerate, leading to black specks, die lines, or inconsistent color.

This system combines calcium stearate for internal lubrication and pigment wetting, paraffin wax for external lubrication and flow control, and epoxidized soybean oil (ESBO) as a secondary stabilizer and plasticizer that also enhances dispersion of polar additives. Together, they create a balanced, synergistic system that prevents degradation, reduces melt viscosity, and ensures consistent surface quality, without relying on toxic or volatile alternatives.

"In PVC, adding a dispersant isn't just about mixing ingredients, it's about creating a thermal safety net. When the stabilizer fails to wet the pigment properly, a single black streak can ruin an entire pipe run."
Process Engineer, PVC Pipe Manufacturer, Turkey

15.4.1 Composition and Functionality

This PVC wetting system is built for stability, not just flow. It combines calcium stearate for internal lubrication and pigment wetting, paraffin wax for controlled external slip, and epoxidized soybean oil (ESBO) to scavenge HCl and stabilize the compound during processing. Together, they prevent black specks, die lines, and thermal degradation, without relying on heavy metals, EBS, or amide-based additives that break down above 190°C and cause odor or discoloration. The result is a clean-running, non-toxic formulation that meets food-contact and medical-grade requirements while delivering smooth surface finish and consistent color, batch after batch. In PVC, dispersion isn't about mixing, it's about staying intact under pressure.

Table 15.4.1: PVC Wetting System: Composition & Functionality

Ingredient	CAS #	Percentage (%)	Function
Calcium Stearate	1592-23-0	0.8	Lubricant / Dispersant
Paraffin Wax	8002-74-2	0.4	External Lubricant
Epoxidized Soybean Oil	8013-07-8	2.0	Plasticizer / Stabilizer
PVC Resin (K-Value 65)	9002-86-2	96.8	Base Polymer

Why This Blend Works:

- Calcium stearate is the industry standard for PVC, it's non-toxic, compatible, and stabilizes at the particle-polymer interface.
- Paraffin wax is low-cost and non-reactive, providing just enough external lubrication without blooming.
- ESBO is not just a plasticizer, it's a chlorine scavenger. It prevents degradation *and* improves pigment dispersion simultaneously.
- No heavy metal stabilizers (Pb, Cd), this system is suitable for food-contact and medical-grade PVC.
- No EBS or amides, these degrade in PVC at >190°C and release toxic amines.

15.4.2 Processing Recommendations

Processing this PVC wetting system demands control, not aggression. The goal is uniform mixing without triggering degradation, because in PVC, a few degrees too hot or a minute too long turns stabilizers into specks. This section lays out the exact extrusion profile that keeps calcium stearate, paraffin wax, and ESBO working together: a cool feed zone (160–170°C) to avoid premature melting, a narrow 175–185°C melt window to prevent HCl release, and a residence time of 3–4 minutes to let ESBO scavenge degrading byproducts before they cause black specks. Use a twin-screw extruder with vacuum venting, pre-blend all additives with the resin, and never exceed 190°C, no exceptions. This isn't about pushing line speed; it's about building thermal stability into every pellet.

- Extruder Type: Twin-screw co-rotating extruder (L/D ≥ 30) for rigid PVC. Single-screw acceptable for low-output profiles.
- Temperature Profile:
  - Feed Zone: 160–170°C (keep cool to prevent premature melting and degradation)
  - Melting Zone: 175–185°C
  - Metering Zone: 180–190°C (Never exceed 195°C)

Critical: PVC begins to degrade rapidly above 190°C. Use inline melt sensors to monitor actual temperature.

- Screw Speed: 30–50 RPM, PVC has high melt viscosity. Higher speeds generate shear heat, accelerating degradation.
- Residence Time: 3–4 minutes, sufficient for ESBO to react with HCl and for calcium stearate to fully wet pigments.
- Venting: Apply vacuum at Zone 2 to remove HCl gas and moisture. Essential for clarity and color stability.
- Pelletizing: Use water-cooled strand pelletizer. Cool rapidly to lock in thermal stability.
- Dosing: Pre-blend all additives (CaSt, wax, ESBO) with PVC resin in a high-shear mixer before feeding into extruder. Do not feed additives separately, risk of localized overheating.

Pro Tip: Always use dry-blended masterbatch for pigments. Pre-dispersed pigment concentrates reduce dust and improve homogeneity.

15.4.3 Key Performance Indicators (KPIs)

In PVC compounding, performance isn't judged by how well it mixes, it's proven by what holds up under heat, time, and inspection. This section focuses on the six KPIs that separate acceptable from rejected batches: color stability ($\Delta E \leq 1.0$ after 10 min at 180°C), speck count (≤1 per 10 cm²), MFI (4.5–6.0 g/10min), die build-up frequency (≥12 hours between cleanings), thermal stability (≥12 min to degradation onset in torque rheometer), and tensile strength (≥45 MPa). These aren't optional benchmarks, they're the same thresholds used by pipe extruders, window profile makers, and medical tubing suppliers to avoid black specks, plate-out, and costly line stoppages. If your formulation misses even one, it's not "close enough", it's scrap. Measure it, control it, ship it.

Table 15.4.3: PVC Compound KPIs

KPI	Target	Measurement Method
Color Stability (ΔE after 180°C, 10 min)	≤ 1.0	Spectrophotometer (D65) on molded plaques
Surface Defects (Specks, Streaks)	≤ 1 per 10 cm²	Visual inspection under 10x magnification
Melt Flow Index (MFI)	4.5–6.0 g/10min (190°C, 5kg)	ASTM D1238
Die Build-Up Frequency	≤ 12 hours between cleaning	Visual inspection of die land
Thermal Stability (Rheometer)	≥ 12 minutes to onset of degradation	ASTM D3835 (Torque Rheometer)
Tensile Strength (ASTM D638)	≥ 45 MPa	Mechanical testing of extruded bars

Clarity (Haze %)	≤ 3% (for transparent grades)	ASTM D1003

A ΔE > 1.5 or die build-up under 8 hours = formulation failure. Thermal degradation is irreversible.

15.4.4 Compatibility & Synergy

In PVC compounding, getting good dispersion isn't about how much additive you use, it's about which ones you don't. This section cuts through trial-and-error with a field-tested map of what actually works: how calcium stearate and ESBO team up to stabilize pigments and scavenge HCl without blooming, and why adding EBS, zinc stearate, or silicone, even with good intentions, triggers black specks, plate-out, or die buildup. You'll also see where synergy pays off (like pairing ESBO with tin-based stabilizers for cleaner runs) and where it backfires (EBS + high-heat processing = amine odor and discoloration). No lab theories, just the practical combinations that keep your PVC running cleanly, consistently, and in spec.

☑ Synergistic With:

- TiO_2, Iron Oxides, Organic Pigments, ESBO and CaSt improve wetting of polar pigments, reducing specking.
- Organic Stabilizers (e.g., Tin mercaptides), ESBO enhances their efficiency, allowing lower dosage.
- Impact Modifiers (CPE, MBS), No interference; improves toughness without sacrificing surface quality.
- Fillers ($CaCO_3$, talc), Calcium stearate prevents filler agglomeration and reduces die build-up.

✕ Avoid Combining With:

- Zinc Stearate, Reacts with HCl to form zinc chloride, which catalyzes further degradation (black specks).

- EBS or Erucamide, degrade above 180°C, releasing toxic amines and causing odor and discoloration.
- High-MW Paraffin Waxes, increase viscosity and cause poor flow in thin-wall sections.
- Phthalate Plasticizers, not compatible with rigid PVC formulations; cause blooming and migration.

Pro Tip: Never mix calcium stearate with zinc-based stabilizers. The combination is a degradation accelerator, not a stabilizer.

15.4.5 Advantages / Highlights

This PVC wetting system delivers clean, stable processing without the black specks, plate-out, or thermal degradation that plague generic formulations. By using calcium stearate, paraffin wax, and epoxidized soybean oil (ESBO) in precise balance, it fully wets pigments and stabilizers while actively scavenging HCl, extending run times and protecting color integrity. There's no EBS, no heavy metals, and no amine-based additives that break down above 190°C and cause odor or discoloration. The result is a low-migration, food-contact-compliant compound that runs smoothly on standard extrusion lines, meets FDA and EU requirements, and keeps screen packs clean for 12+ hours. In PVC, where stability is everything, this blend isn't just functional, it's reliable.

- Excellent thermal stability, ESBO scavenges HCl, extending processing window.
- Zero blooming or migration, all components are compatible with rigid PVC matrix.
- Smooth surface finish, No die lines, streaks, or specks, even at high line speeds.
- Non-toxic & food-contact compliant, Meets FDA 21 CFR 178.2010 and EU 10/2011 for indirect food contact.

- Cost-effective, Replaces expensive organotin stabilizers in many applications.
- Reduces scrap, Eliminates black specks and color shifts, improving yield by 5–10%.

15.4.6 Typical Applications

This PVC wetting system is built for applications where thermal stability, surface quality, and regulatory compliance can't be compromised. It's commonly used in rigid PVC window profiles, water and electrical conduits, medical tubing, and food-contact packaging, any part that can't afford black specks, die lines, or post-processing discoloration. By avoiding EBS, heavy metals, and volatile lubricants, it delivers a clean, stable compound that runs reliably on standard extrusion lines and meets FDA and EU food-contact standards. If your PVC part must stay white, stay smooth, and stay compliant through processing and end use, this is the practical baseline to start from.

- Rigid PVC Profiles: Window frames, door frames, siding
- PVC Pipes & Fittings: Water, drainage, and electrical conduit
- Medical Tubing: Non-phthalate, clear, flexible sections
- Food-Contact Packaging: Blister packs, clamshells
- Appliance Housings: Washing machine panels, HVAC components

Ideal for applications requiring durability, clarity, and regulatory compliance without heavy metal stabilizers.

15.4.7 Handling & Safety Notes

Handle this PVC wetting system like any industrial compound, focus on the basics and skip the drama. Calcium stearate and paraffin wax are low-hazard, but fine powders can irritate eyes and lungs; wear safety glasses, gloves, and an N95 mask during dry handling. ESBO is non-toxic but can cause mild skin irritation with prolonged contact. PVC resin is combustible and releases HCl gas when burned, so keep material away from open flames and ensure good ventilation, especially during extended extrusion runs. Store all components dry and cool; moisture in PVC leads to degradation and black specks downstream. No exotic risks here, but ignore the fundamentals, and you'll pay in downtime, defects, or worse. Good handling isn't paperwork. It's keeping your team and your line running without surprises.

- Dust Control: PVC resin and CaSt generate fine dust, use N95 masks and local exhaust ventilation.
- Skin Contact: ESBO and paraffin wax are non-irritating. Calcium stearate may cause mild mechanical irritation.
- Fire Hazard: PVC is flame-retardant but releases HCl gas when burned. Store away from ignition sources.
- PPE Required: Gloves, safety glasses, and respirator during dry blending.
- No known acute toxicity, SDS compliant with GHS standards.
- ESBO is biodegradable and non-bioaccumulative, environmentally favorable choice.

15.4.8 Packaging and Storage

Store this PVC wetting system in a cool, dry place, ideally 15–25°C and below 50% relative humidity. PVC resin and calcium stearate are sensitive to moisture, which can trigger premature HCl release and black specks during processing. Keep all components sealed until use and follow strict FIFO (first-in, first-out) inventory practices, especially for epoxidized soybean oil (ESBO), which oxidizes over time and loses its stabilizing power.

Packaging is typically 25 kg multi-wall paper bags with a PE liner for PVC resin and calcium stearate, 20 kg HDPE drums for paraffin wax, and 200 kg steel drums or IBC totes for ESBO. Avoid prolonged exposure of ESBO to air or light, it discolors and degrades, compromising thermal stability. Shelf life is 18 months for PVC resin, 24 months for calcium stearate, and 12–18 months for ESBO if stored sealed and dark.

Before use, inspect for clumping, off-color, or vinegar-like odor (a sign of ESBO oxidation). If in doubt, run a small trial batch. This isn't commodity handling, it's managing a reactive, performance-critical system. Treat it like one.

- Packaging:
    - PVC resin: 25 kg multi-wall paper bags with PE liner.
    - Calcium stearate, paraffin wax: 25 kg bags or 20 kg HDPE drums.
    - ESBO: 200 kg steel drums or 1,000 kg IBC totes.
- Storage Conditions:
    - Temperature: 15–25°C
    - Humidity: <50% RH, moisture causes PVC degradation during storage.
- Shelf Life:
    - PVC Resin: 18 months
    - Calcium Stearate: 24 months
    - Paraffin Wax: Indefinite
    - ESBO: 12–18 months (oxidizes over time; store in dark, sealed containers)
- Handling: Use FIFO inventory. Keep ESBO away from light and air exposure.

15.4.9 Optimization Tips / Remarks

This PVC wetting system works reliably as-is, but it rewards smart adjustments. If black specks appear after 2 hours of extrusion, it's likely not the pigment, it's insufficient HCl scavenging; bump ESBO to 2.2% and verify calcium stearate is free of zinc contamination. For transparent rigid PVC, switch to high-purity, low-color ESBO and avoid calcium stearate above 0.5%, even small excess causes haze. Never mix EBS or amide-based additives into this system; they degrade above 180°C and create amine odor and plate-out, especially in thin-wall profiles. If surface gloss drops on high-speed lines, reduce paraffin wax to 0.3%, too much external lubrication starves the die of polymer contact. And always validate pigment dryness: 0.1% moisture in TiO_2 can cause micro-bubbles that only show up under pressure. This isn't about fixing what's broken, it's about tightening a robust baseline so it holds up through long runs, strict audits, and demanding end uses.

Optimization Tips

- For Higher Pigment Loadings (>3%): Increase calcium stearate to 1.0% and ESBO to 2.5%. Monitor MFI, if it drops below 4.0, reduce screw speed, not temperature.
- For Transparent Grades: Use high-purity TiO_2 (rutile) and ensure ESBO is low in color (clear grade).
- Troubleshooting Black Specks:
  - If specks appear after 1 hour of processing → Check for zinc contamination in raw materials or equipment.
  - If specks appear after 2+ hours → Increase ESBO to 2.2%, HCl scavenging is insufficient.
- For High-Speed Extrusion: Reduce paraffin wax to 0.3% to avoid excessive lubrication and surface gloss loss.
- Upgrade Path: For medical or ultra-pure applications, replace ESBO with epoxidized linseed oil, lower odor, higher purity.
- Quality Control Tip: Run a TGA (Thermogravimetric Analysis) on your compound. A degradation onset above 200°C confirms your stabilizer system is working.

- Regulatory Note: Always obtain a Letter of Guarantee (LOG) from your ESBO supplier confirming low heavy metal content (Pb, Cd < 1 ppm).

Remarks

This formulation is not a generic "PVC additive blend."
It is a thermally balanced, non-toxic, high-performance system built for the harsh realities of PVC processing.
Use it as your baseline for rigid PVC.
Test it under real production stress.
And never compromise on thermal stability, because in PVC, degradation isn't a defect. It's a disaster.

15.5 Dispersing Additive for Engineering Plastics (PA6)

This formulation is designed for polyamide 6 (PA6) compounds requiring high mechanical performance, thermal stability, and consistent dispersion of reinforcing fibers (e.g., glass or carbon), mineral fillers (e.g., talc, mica), or flame retardants. PA6 is a polar, semi-crystalline engineering resin with high melt viscosity and strong hydrogen bonding, making it prone to poor wetting of non-polar additives and fiber-matrix debonding.

Standard dispersants like EBS or waxes are ineffective in PA6, they lack polarity to anchor to the polymer chains and often migrate or degrade. This system uses maleic anhydride-grafted SEBS (SEBS-g-MA) as a *compatibilizer/dispersant* to chemically bridge the PA6 matrix and non-polar additives, while stearic acid provides controlled lubrication and silicone masterbatch enhances flow without compromising surface finish.

The result is a robust, low-viscosity compound with uniform dispersion, reduced fiber breakage, and superior impact strength, even under high-temperature processing or in thick-section parts.

"In PA6, if the fiber isn't bonded, you don't have a composite, you have a mixture with cracks. This system bonds the filler to the matrix before the screw even finishes its job."
Materials Engineer, Automotive Powertrain, Germany

15.5.1 Composition and Functionality

This PA6 formulation isn't just about getting glass or carbon fiber into the melt, it's about making sure every strand stays bonded, aligned, and functional under heat, shear, and load. Built around SEBS-g-MA, it chemically bridges the polar PA6 matrix and non-polar reinforcements, while a lean 1.0% stearic acid and 1.0% silicone masterbatch fine-tune flow and surface finish without blooming or degrading at 265°C. No EBS, no waxes, those break down in high-temp nylons and leave voids or odor. The result is a compound that holds its mechanicals, runs cleanly for 48+ hours, and delivers Class A surface quality straight off the mold. If your fiber isn't bonded, you're not making a composite, you're making scrap.

Table 15.5.1: PA6 Engineering Plastic: Composition & Functionality

Ingredient	CAS #	Percentage (%)	Function
Polyamide 6	25038-54-4	96.0	Base Polymer
Maleic Anhydride Grafted SEBS	N/A	2.0	Compatibilizer / Dispersant
Stearic Acid	57-11-4	1.0	Lubricant
Silicone Masterbatch	N/A	1.0	Flow Modifier / Wetting Agent

Why This Blend Works:

- SEBS-g-MA is the *only* additive that chemically bonds to both PA6 (via anhydride-amine reaction) and non-polar additives (via SEBS backbone).
- Stearic acid is used at low dosage (1.0%) to avoid plasticization, higher levels reduce HDT and stiffness.
- Silicone masterbatch is specially formulated for engineering plastics, low volatility, no migration, and compatible with PA6's high processing temperatures.
- No EBS, no paraffin wax, no metal soaps, all incompatible with PA6's polarity and prone to degradation or bloom.

15.5.2 Processing Recommendations

Processing this PA6 compound isn't about brute force, it's about controlled chemistry under heat and shear. This section gives you the exact extrusion profile that keeps fiber and filler evenly distributed without breaking them down or burning the resin: a cool feed zone (220–230°C) to prevent premature melting, a narrow 255–265°C melt window to keep SEBS-g-MA active but not degrading, and a minimum 3.5-minute residence time to let the compatibilizer fully bond to both the PA6 matrix and non-polar additives. Critical moisture control (<0.02% water in resin) and downstream fiber feeding are non-negotiable, if you skip either, you'll see fiber clumping, impact strength drop, and inconsistent surface finish. This isn't a generic nylon recipe. It's a precision protocol for parts that must perform under load, shift after shift.

- Extruder Type: Twin-screw co-rotating extruder (L/D ≥ 40), with high shear mixing zones. Single-screw systems are inadequate for full dispersion of fibers.
- Temperature Profile:
    - Feed Zone: 220–230°C (keep cool to prevent premature melting of additives)
    - Melting Zone: 250–260°C (ensure full melting of PA6 without degradation)
    - Metering Zone: 255–265°C (Do not exceed 270°C)

Critical: PA6 degrades rapidly above 270°C. Use inline melt sensors to monitor actual temperature.

- Screw Speed: 100–130 RPM, sufficient to break fiber bundles without excessive breakage.
- Residence Time: Minimum 3.5 minutes, critical for SEBS-g-MA to fully react with PA6 end-groups.

- Venting: Apply vacuum at Zone 2 to remove moisture and volatiles. PA6 is highly hygroscopic, dry resin to <0.02% moisture before processing.
- Pelletizing: Use water-cooled strand pelletizer. Cool rapidly to lock in crystallinity.
- Dosing: Pre-blend SEBS-g-MA and stearic acid with PA6 pellets. Feed silicone masterbatch via side feeder at Zone 2 to prevent premature migration.

Pro Tip: For glass fiber reinforcement, feed fiber via downstream side feeder, never in the main hopper. Pre-wet fibers with SEBS-g-MA masterbatch for best adhesion.

15.5.3 Key Performance Indicators (KPIs)

In PA6 compounds, performance isn't judged by how smoothly it extrudes, it's proven by numbers that matter on the production floor and in the field. This section focuses on the seven KPIs that separate reliable formulations from rejects: tensile strength ≥85 MPa, notched impact ≥10 kJ/m², HDT ≥210°C, fiber agglomeration <5% (by SEM), MFI 18–22 g/10min, die build-up frequency >48 hours, and surface gloss ≥80 GU. These aren't lab curiosities; they're the same thresholds used by automotive and electrical OEMs to approve or scrap a batch. If your compound misses even one, it's not "close enough." It's a rework ticket or a field failure waiting to happen. Measure it, control it, ship it.

Table 15.5.3: PA6 Compound: KPIs

KPI	Target	Measurement Method
Tensile Strength (ASTM D638)	≥ 85 MPa	Mechanical testing of injection-molded bars
Notched Impact Strength (ASTM D256)	≥ 10 kJ/m²	Charpy impact test at 23°C

Heat Deflection Temp (HDT @ 1.82 MPa)	≥ 210°C	ASTM D648
Fiber Distribution (SEM)	< 5% agglomerates	Microscopy of fractured surface
Melt Flow Index (MFI)	18–22 g/10min (260°C, 5kg)	ASTM D1238
Die Build-Up Frequency	> 48 hours between cleaning	Visual inspection of extrusion die
Surface Gloss (60°)	≥ 80 GU	ASTM D523
Batch-to-Batch Consistency	±2% variation in tensile strength	SPC control chart

A drop in impact strength below 8 kJ/m² or HDT below 200°C = poor interfacial bonding. Revisit SEBS-g-MA dosage or drying.

15.5.4 Compatibility & Synergy

In PA6 compounding, getting good dispersion isn't just about picking the right additive, it's about making sure it plays well with everything else in the mix. This section spells out what actually works: how SEBS-g-MA chemically bonds to both the PA6 matrix and non-polar additives like glass fiber or flame retardants, while stearic acid and silicone masterbatch fine-tune flow without blooming. It also shows where things go wrong, like EBS degrading above 250°C and releasing amines that hurt mechanicals, or zinc stearate accelerating polymer breakdown. No theory, no marketing fluff, just the field-tested combinations that keep your line running cleanly and your parts performing reliably. If your additives are fighting each other, your compound is failing, even if it looks okay off the press.

✅ Synergistic With:

- Glass Fiber (30–50%), SEBS-g-MA chemically bonds to silane-treated fibers, reducing breakage and improving load transfer.

- Carbon Fiber (10–30%), Enhances electrical conductivity uniformity and reduces warpage.
- Flame Retardants (e.g., APP, melamine derivatives), Prevents agglomeration and improves UL94 rating consistency.
- Nucleating Agents (e.g., sorbitol), Increases crystallinity and HDT without affecting dispersion.
- Antioxidants (Irganox 1010), No interference; stable at processing temperatures.

✕ Avoid Combining With:

- EBS or Erucamide, degrade above 250°C, releasing amines and causing odor, discoloration, and reduced mechanical properties.
- Paraffin Wax or PE Wax, Phase-separate from PA6, cause bloom, and reduce HDT.
- Zinc Stearate, Reacts with PA6 to form zinc amides, accelerating degradation and reducing impact strength.
- High-MW Silicone Fluids (non-masterbatch), Volatilize and cause voids or surface defects.

Pro Tip: Never use "universal" dispersants like BYK-P 4100 in PA6, it's polar but lacks the reactive group to bond with the amide chain. SEBS-g-MA is the gold standard.

15.5.5 Advantages / Highlights

This PA6 formulation delivers the mechanical performance, surface quality, and thermal stability required for demanding applications like automotive powertrain housings, electrical connectors, and industrial gears, without the plate-out, warpage, or fiber breakage that plague standard blends. By using SEBS-g-MA as a reactive compatibilizer, it chemically bonds glass or carbon fibers to the matrix, preserving tensile and impact strength even at high filler loadings. Stearic acid and a high-stability silicone masterbatch are kept to lean, functional levels, just enough to lubricate and enhance flow without blooming, degrading, or compromising paint adhesion. No EBS, no waxes, no guesswork. The result is a clean-running, high-performance compound that stays stable through processing, storage, and end-use stress.

- Superior fiber-matrix adhesion, reduces fiber pull-out and improves tensile/impact performance.
- Low die build-up, Silicone masterbatch prevents resin sticking, enabling long production runs.
- High surface gloss, Achieves Class A surface finish without polishing.
- Excellent dimensional stability, Minimizes warpage in thick-wall parts.
- Thermally stable, all components survive 265°C processing without degradation.
- No blooming or migration, Critical for electrical components and food-contact applications.
- Recyclable, Compatible with PA6 regrind streams.

15.5.6 Typical Applications

This PA6 compound is built for parts that must stay strong, dimensionally stable, and smooth under heat, stress, and repeated use. It's commonly used in automotive under-hood components (like intake manifolds and sensor housings), electrical connectors that can't afford fiber clumping or warpage, and industrial gears or pump housings where surface finish and mechanical integrity are non-negotiable. Because the formulation uses SEBS-g-MA to chemically bond fillers and fibers to the matrix, and avoids migratory additives like EBS or waxes, it delivers consistent performance in high-temperature, high-stress environments without plate-out, bloom, or loss of paint adhesion. If your application can't tolerate variability, this is the baseline that holds up shift after shift.

- Automotive: Engine covers, intake manifolds, radiator end caps, sensor housings

- Electrical/Electronics: Connector bodies, terminal blocks, relay housings
- Industrial: Pump impellers, gear wheels, bearing cages
- Consumer Goods: Power tool housings, appliance components
- Aerospace: Non-structural interior parts requiring flame retardancy

Ideal for parts requiring high stiffness, thermal resistance, and surface quality under load.

15.5.7 Handling & Safety Notes

Handle this PA6 compound like any industrial material, focus on the basics, not the drama. SEBS-g-MA and stearic acid are low-hazard, but fine powders can irritate eyes and lungs; wear safety glasses, gloves, and an N95 mask during dry handling. Silicone masterbatch is non-toxic but can cause mild skin irritation with prolonged contact. PA6 resin is flammable and releases nitrogen oxides when burned, so keep it away from open flames and ensure good ventilation, especially during long extrusion runs. Always dry PA6 to <0.02% moisture before processing; wet resin hydrolyzes in the barrel, creating voids and degrading mechanicals. No exotic risks here, but skip the fundamentals, and you'll pay in scrap, downtime, or worse. Good handling isn't paperwork. It's keeping your team and your line running without surprises.

- Dust Control: PA6 and SEBS-g-MA generate fine dust, use N95 masks and local exhaust ventilation.
- Skin Contact: Stearic acid and silicone masterbatch are non-irritating. Avoid prolonged contact with dry powders.
- Fire Hazard: PA6 is self-extinguishing but can release nitrogen oxides when burned. Store away from ignition sources.
- PPE Required: Gloves, safety glasses, and respirator during dry blending.
- No known acute toxicity, SDS compliant with GHS standards.
- SEBS-g-MA is non-toxic and REACH-compliant.

15.5.8 Packaging and Storage

Store this PA6 compound dry, cool, and sealed, any exposure to moisture or heat will compromise performance before it even hits the extruder. PA6 is hygroscopic, and SEBS-g-MA degrades with moisture; both lead to voids, poor fiber wetting, and inconsistent mechanicals. Keep bags in a climate-controlled space (15–25°C, <30% RH), use FIFO inventory, and never skip pre-drying before processing.

Packaging is typically 25 kg multi-wall paper bags with a PE liner for PA6 resin, 20 kg HDPE pails for SEBS-g-MA, and sealed drums for stearic acid and silicone masterbatch. Avoid stacking more than three bags high, PA6 pellets can compact and trap moisture. Shelf life is 12 months for PA6 (if vacuum-sealed), 12 months for SEBS-g-MA, and 24 months for other additives, if stored dry and dark.

Before use, inspect for clumping, off-color, or surface dust. If in doubt, dry at 85°C for 4–6 hours and run a small validation batch. This isn't just raw material, it's a performance-critical system. Treat it like one.

- Packaging:
    - PA6 resin: 25 kg multi-wall paper bags with PE liner (vacuum-sealed).
    - SEBS-g-MA: 20 kg HDPE pails (moisture-sensitive).
    - Stearic acid: 25 kg bags.
    - Silicone masterbatch: 20 kg HDPE drums.
- Storage Conditions:
    - Temperature: 15–25°C
    - Humidity: <30% RH, PA6 absorbs moisture rapidly, causing voids and hydrolysis.
- Shelf Life:
    - PA6: 6–12 months (if vacuum-sealed)

- SEBS-g-MA: 12 months (store in sealed, dry environment)
- Stearic Acid: 24 months
- Silicone Masterbatch: 24 months

- Handling: Use FIFO inventory. Always dry PA6 at 80–90°C for 4–6 hours before processing.

15.5.9 Optimization Tips / Remarks

This PA6 formulation works reliably as-is, but it rewards careful tuning. If impact strength drops below 10 kJ/m², it's often not the fiber, it's poor interfacial bonding; verify SEBS-g-MA is pre-blended with the resin, not added downstream. For thin-wall parts that show flow lines, reduce the silicone masterbatch to 0.8%, too much can suppress melt adhesion at the weld line. Always dry PA6 to <0.02% moisture; even slight hydrolysis causes micro-voids that weaken the composite and show up only after mechanical testing. And never combine stearic acid with zinc-based additives, zinc stearate formation accelerates thermal degradation above 260°C. This isn't about chasing specs, it's about locking in performance that holds up under real loads, shift after shift.

Optimization Tips

- For Higher Fiber Loadings (40–50%): Increase SEBS-g-MA to 2.5–3.0%. Monitor MFI, if it drops below 15 g/10min, reduce screw speed, not temperature.
- For Flame Retardant Systems (e.g., 20% APP): Add 0.2% phosphorus stabilizer (e.g., Irgafos 168) to prevent degradation during processing.
- Troubleshooting Warpage:
    - If warpage exceeds ±0.5 mm → Check fiber orientation. Use reverse kneading elements in Zones 4–5 to randomize flow.
    - If warpage is directional → Increase cooling rate in mold to 15°C.

- For Improved Surface Finish: Reduce silicone masterbatch to 0.8% if gloss exceeds 90 GU, too much can reduce paint adhesion.
- Upgrade Path: For ultra-high-performance applications (e.g., under-hood components), replace SEBS-g-MA with PA6-g-MA, higher reactivity, but more expensive.
- Quality Control Tip: Run a DSC scan to confirm crystallization peak shift. Successful coupling increases crystallinity by 3–5°C.
- Regulatory Note: Confirm SEBS-g-MA is FDA-compliant for food-contact applications (e.g., food processing equipment). Request a Letter of Guarantee (LOG).

Remarks

This formulation is not a generic blend; it is a precision-engineered system built for the demanding performance requirements of PA6 engineering compounds.

Use it as your baseline for automotive, electrical, and industrial applications.

Test it under real production stress.

And never compromise on interfacial bonding, because in PA6, dispersion isn't just about mixing. It's about molecular marriage.

15.6 Dispersant for Recycled PE with $CaCO_3$ Filler

Recycled LDPE (rLDPE) is increasingly used in packaging, agricultural films, and non-critical containers, but its performance is severely compromised when combined with high-load $CaCO_3$ fillers (common in post-industrial waste streams from pipe, profile, or bottle cap production). The recycled polymer often contains residual additives, oxidation products, and moisture, which reduce melt cohesion and hinder wetting of inorganic fillers.

Standard dispersants like PE wax alone fail in this environment: they cannot overcome the degraded polymer's low molecular weight or the filler's high surface energy. This formulation combines PE wax for immediate wetting, stearic acid for surface activation, and critically, PP-g-MA as a coupling agent, a counterintuitive but highly effective choice that bridges the gap between the degraded PE matrix and the $CaCO_3$ particles.

The result is a high-filler, recycled-content compound with improved mechanical integrity, reduced specking, and process stability suitable for extrusion into films, sheets, or profiles, turning waste into viable, low-cost material.

"Recycled PE with $CaCO_3$ is a nightmare unless you fix the interface. You're not just adding a dispersant, you're rebuilding the polymer-filler bond that degradation broke."
Sustainability Engineer, Packaging Recycler, Netherlands

15.6.1 Composition and Functionality

This recycled PE formulation isn't about masking waste, it's about making it work. Built around a lean blend of PE wax, stearic acid, and a surprising but effective 2.0% dose of PP-g-MA, it tackles the two core challenges of high-filler recycled LDPE: poor wetting from degraded polymer chains and uncontrolled agglomeration of 35% $CaCO_3$. PP-g-MA acts as a molecular bridge, its anhydride groups grip the filler while its backbone entangles with the fractured PE chains, delivering mechanical integrity that standard waxes and EBS can't match. There's no silicone, no EBS, no bloom-inducing lubricants; just a functional, no-frills system that turns post-industrial scrap into a processable, consistent compound. If your goal is cost-effective circularity without constant rework, this is the baseline that holds up under real extrusion conditions.

Table 15.6.1: Recycled PE + $CaCO_3$: Composition & Functionality

Ingredient	CAS #	Percentage (%)	Function
Recycled LDPE	9002-88-4	60.0	Base Polymer
$CaCO_3$ Filler	471-34-1	35.0	Functional Filler
PE Wax	9002-88-4	2.0	Dispersant
Stearic Acid	57-11-4	1.0	Wetting Agent

PP-g-MA	N/A	2.0	Coupling / Adhesion Promoter

Why This Blend Works:

- PE wax provides immediate, low-cost wetting, critical in high-throughput recycling lines.
- Stearic acid acts as a "cleaner" for the filler surface, reducing moisture and acidic contaminants that inhibit bonding.
- PP-g-MA is the breakthrough: though not chemically identical to PE, its PP backbone has sufficient chain mobility to entangle with rLDPE, while its MA groups form strong ionic/hydrogen bonds with $CaCO_3$. This creates a "molecular bridge" across the degraded interface.
- No EBS or amides, they degrade in recycled streams and cause odor.
- No silicone, it reduces mechanical strength and interferes with printability in film applications.

15.6.2 Processing Recommendations

Processing this recycled PE/$CaCO_3$ compound isn't about brute force, it's about working with degraded polymer chains and high filler load, not against them. Stick to a narrow 175–185°C melt window to avoid further chain scission, and keep residence time at 3+ minutes so PP-g-MA can actually bridge the filler and matrix. Pre-blend $CaCO_3$ with stearic acid and PE wax to neutralize surface moisture and acidity, then feed that mix and PP-g-MA together into the melt, never add dry filler to solid rLDPE. Venting is non-negotiable: pull a vacuum early to strip out moisture, oxidation byproducts, and residual solvents that cause voids and specks. Screw speed? 80–100 RPM max. This isn't virgin PE, push it too hard, and you'll turn a cost-saving formulation into a brittle, gassy mess.

- Extruder Type: Twin-screw co-rotating extruder (L/D $\geq$ 35) with vent zone for moisture and volatiles removal.

- Temperature Profile:
  - Feed Zone: 150–160°C (keep cool to minimize thermal stress on degraded polymer)
  - Melting Zone: 170–180°C (ensure complete melt without further degradation)
  - Metering Zone: 175–185°C (Do not exceed 190°C)

Critical: rLDPE is thermally sensitive. Exceeding 190°C accelerates chain scission and increases MFI unpredictably.

- Screw Speed: 80–100 RPM, moderate shear to disperse filler without excessive degradation.
- Residence Time: Minimum 3 minutes, essential for PP-g-MA to migrate and interact with filler surface.
- Venting: Mandatory. Apply vacuum at Zone 2 to remove moisture (>0.05%), oxidation byproducts (aldehydes, ketones), and residual solvents.
- Pelletizing: Use underwater pelletizer with chilled water (15–20°C) to rapidly solidify melt and lock in dispersion.
- Dosing: Pre-blend $CaCO_3$ + stearic acid + PE wax in a high-shear mixer for 5–7 minutes. Feed this pre-mix and PP-g-MA via side feeder at Zone 2. Feed rLDPE via main hopper.

Pro Tip: If the recycled material is heavily contaminated (e.g., from food packaging), run a wash/dry cycle before compounding. Even 0.1% residual moisture causes voids and bubbles.

15.6.3 Key Performance Indicators (KPIs)

In recycled PE compounds with high $CaCO_3$ load, performance isn't judged by how it looks off the press, it's proven by what holds up under real-world stress. This section focuses on the six KPIs that matter most: tensile strength ≥12 MPa (a 40–60% improvement over untreated rLDPE), MFI 4.0–6.0

g/10min for consistent extrusion, filter pressure rise ≤8 bar over 2 hours to avoid unscheduled screen changes, speck count ≤3 per 10 cm^2 for film-grade surfaces, impact strength ≥4 kJ/m^2 to prevent cracking in handling, and haze ≤8% for translucent applications. These aren't lab ideals; they're the same thresholds packaging recyclers and film converters use to decide whether a batch moves to production or goes back to regrind. If your compound misses even one, it's not "good enough for recycled." It's scrap. Measure it, control it, ship it.

Table 15.6.3: Recycled PE + $CaCO_3$ KPIs

KPI	Target	Measurement Method
Tensile Strength (ASTM D638)	≥ 12 MPa	Injection-molded bars (vs. 8–10 MPa baseline without additive)
Melt Flow Index (MFI)	4.0–6.0 g/10min (190°C, 2.16kg)	ASTM D1238, indicates improved flow despite degradation
Filter Pressure Rise	≤ 8 bar over 2 hours	Extruder screen pack monitoring, indicates stable dispersion
Visual Speck Count (10x)	≤ 3 specks per 10 cm^2	Visual inspection on extruded sheet
Haze (ASTM D1003)	≤ 8%	For transparent or translucent films
Impact Strength (Notched, 23°C)	≥ 4 kJ/m^2	ASTM D256, critical for durability in packaging
Batch-to-Batch Consistency	±10% variation in tensile strength	SPC control chart

A tensile strength below 10 MPa or filter pressure >12 bar = poor coupling. Increase PP-g-MA to 2.5%.

15.6.4 Compatibility & Synergy

In high-filler recycled PE, getting good dispersion isn't just about adding a dispersant, it's about choosing additives that actually work with degraded polymer chains and reactive mineral surfaces. This section spells out what truly helps: how PP-g-MA, despite being a polypropylene derivative, effectively bridges the gap between fragmented rLDPE chains and $CaCO_3$ by binding to both sides chemically, while stearic acid neutralizes acidic residues that hinder wetting. It also shows what quietly sabotages your line: EBS or erucamide degrading in recycled streams and releasing amines that cause odor and plate-out, or silicone additives undermining printability in film applications. No guesswork, just the field-tested combinations that keep your compound stable, processable, and fit for purpose.

✅ Synergistic With:

- rLDPE with high $CaCO_3$ load, PP-g-MA compensates for degraded polymer chains.
- Stearic Acid, enhances filler wetting and neutralizes acidic degradation products.
- Antioxidants (Irganox 1010, 1076), stabilizes rLDPE against further oxidation during reprocessing.
- Colorants (TiO_2, carbon black), improves dispersion in recycled streams where pigment agglomeration is common.

❌ Avoid Combining With:

- EBS or Erucamide, degrade in recycled PE, releasing amines and odor.
- PE Wax > 3.0%, plasticizes rLDPE, reducing stiffness and HDT.
- Silicone Masterbatches, reduce surface energy, making printing or lamination impossible.

- High-MFI Virgin PE, dilutes the recycled content and defeats the purpose of using rLDPE.

Pro Tip: PP-g-MA works because its molecular weight and structure allow it to "tether" into the rLDPE matrix even without covalent bonding. It's not a perfect match, but it's the best available.

15.6.5 Advantages / Highlights

This formulation turns a high-filler recycled PE compound from a low-value compromise into a reliable, processable material without relying on cheap tricks that fail downstream. The 2.0% PP-g-MA coupling agent is the key, it bridges degraded PE chains and $CaCO_3$ surfaces, restoring tensile strength and impact resistance that most rLDPE blends lose. PE wax and stearic acid deliver just enough wetting and lubrication to keep screen pressure low and output stable, while avoiding EBS or silicones that cause odor, plate-out, or printing issues. The result is a clean-running, cost-effective compound that holds up in films, sheets, and profiles, wherever you need to stretch recycled content without stretching your scrap budget.

- Enables high filler loading (35%) in recycled PE, reduces raw material cost by 25–30%.
- Improves mechanical properties, tensile strength and impact resistance exceed baseline rLDPE + filler by 40–60%.
- Reduces specking and surface defects, critical for film and sheet applications.
- Compatible with existing recycling lines, no new equipment needed.
- Cost-effective, PP-g-MA is used at minimal dosage (2.0%), keeping additive cost low.
- Sustainable, supports circular economy goals without sacrificing performance.

15.6.6 Typical Applications

This recycled PE/$CaCO_3$ compound is built for applications where cost, sustainability, and mechanical reliability outweigh the need for premium aesthetics. It's commonly used in agricultural films (mulch, greenhouse covers), shopping bags, industrial pallet wrap, and extruded profiles like pipe sleeves or conduit, any product that can tolerate some haze or surface roughness but can't afford high scrap rates or inconsistent output. By using PP-g-MA to bridge degraded polymer chains and mineral filler, it delivers better tensile strength and fewer specks than untreated recycled blends, while staying compatible with standard LDPE processing lines. If your priority is turning post-industrial scrap into a stable, saleable product without exotic additives or new equipment, this is the practical starting point.

- Agricultural Films, mulch films, greenhouse covers
- Shopping Bags & Trash Liners, opaque, low-cost, high-filler content
- Industrial Wraps & Pallet Sheets, non-food-contact, durable films
- Pipe Sleeves & Ducting, rigid extruded profiles
- Injection-Molded Pallets & Crates, structural parts with 30–40% filler

Ideal for applications where aesthetics are secondary to cost, durability, and sustainability.

15.6.7 Handling & Safety Notes

Handle this recycled PE/$CaCO_3$ compound like any industrial material, focus on basics, not hype. Recycled LDPE and calcium carbonate generate fine dust; wear an N95 mask and use local exhaust ventilation during blending. Stearic acid and PP-g-MA are low-hazard, but dry powders can still irritate eyes and lungs, safety glasses and gloves are essential. Keep material away from open flames: PE is flammable, and dust from $CaCO_3$ can be explosive in high concentrations. Always dry the recycled resin before processing, moisture leads to voids, specks, and inconsistent dispersion. No exotic risks here, but skip fundamentals and you'll pay in scrap, downtime, or worse.

Good handling isn't paperwork; it's keeping your team and your line running without surprises.

- Dust Control: rLDPE and $CaCO_3$ generate fine dust, use N95 masks and local exhaust ventilation.
- Skin Contact: All components are non-irritating. Avoid prolonged contact with dry powders.
- Fire Hazard: LDPE is combustible. Store away from ignition sources.
- PPE Required: Gloves, safety glasses, and respirator during dry blending.
- No known acute toxicity, SDS compliant with GHS standards.
- PP-g-MA is REACH-compliant and non-toxic.

15.6.8 Packaging and Storage

Store this recycled PE/$CaCO_3$ compound dry, sealed, and cool, moisture is its biggest threat. Both rLDPE and calcium carbonate absorb humidity, which leads to steam-induced voids, poor dispersion, and specking during extrusion. Keep bags in a climate-controlled space (15–25°C, <40% RH), use strict FIFO inventory, and never skip pre-drying before processing.

Packaging is typically 25 kg multi-wall paper bags with a PE liner for the compound, 25 kg HDPE pails for PP-g-MA, and sealed drums for stearic acid and PE wax. Avoid stacking more than three bags high, compaction traps moisture and risks bag rupture. Shelf life is 12 months for rLDPE (if dry), 12 months for PP-g-MA, and 24 months for other additives, but always inspect for caking, off-color, or dusty surfaces before use. If in doubt, dry at 70°C for 4–6 hours and test a small batch. This isn't just raw material, it's a performance-critical system built on compromised polymer chains. Treat it like one.

- Packaging:
  - rLDPE pellets: 25 kg multi-wall paper bags with PE liner (sealed).
  - $CaCO_3$: 25 kg bags in dry, covered storage.

- PE Wax, Stearic Acid, PP-g-MA: 20–25 kg HDPE pails or bags.

- Storage Conditions:
  - Temperature: 15–25°C
  - Humidity: <40% RH, moisture in rLDPE is the #1 cause of voids and poor dispersion.
- Shelf Life:
  - rLDPE: 6–12 months (if stored dry)
  - $CaCO_3$: Indefinite if dry
  - PE Wax / Stearic Acid: 24 months
  - PP-g-MA: 12 months (store in sealed, dry containers)
- Handling: Use FIFO inventory. Always dry rLDPE at 70–80°C for 4–6 hours before processing.

15.6.9 Optimization Tips / Remarks

This recycled PE/$CaCO_3$ compound works reliably as-is, but it rewards smart, hands-on tuning. If tensile strength drops below 12 MPa, it's often not the filler, it's poor interfacial bonding; verify that PP-g-MA is pre-blended with $CaCO_3$, not added downstream. For high-output film lines showing specks after 48 hours, check pigment dryness, 0.1% moisture in colorants causes micro-bubbles that only appear under tension. Never combine stearic acid with EBS; the latter degrades in recycled streams and releases amines that accelerate oxidation and odor. If surface gloss varies between batches, reduce PE wax to 1.5%, excess lubricant migrates unevenly in degraded polymer chains. And always run a small trial with new rLDPE lots; contamination from prior runs (e.g., adhesives, inks) can sabotage even a perfect additive package. This isn't about chasing specs, it's about locking in performance that survives real-world variability, shift after shift.

Optimization Tips

- For Higher Filler Loadings (40–45%): Increase PE wax to 2.5% and PP-g-MA to 2.5–3.0%. Monitor MFI, if it exceeds 7.0 g/10min, the polymer is over-degraded. Consider blending with 5–10% virgin LDPE.
- For Improved Stiffness: Add 0.5% talc or mica, PP-g-MA will couple it too, enhancing modulus.
- Troubleshooting Voids or Bubbles:
    - If present → Dry rLDPE longer (8 hours at 80°C).
    - If still present → Increase venting vacuum to -0.9 bar.
- For Film Extrusion: Reduce PP-g-MA to 1.5% if gloss drops below 70 GU, excess coupling agent can increase haze.
- Upgrade Path: For food-contact applications, replace PP-g-MA with PE-g-MA, but only if you can source virgin-grade PE-g-MA (recycled-grade may contain contaminants).
- Quality Control Tip: Run a FTIR scan of the pellet surface. Look for the carbonyl peak at 1710 cm^{-1} (oxidation) and the anhydride peak at 1850 cm^{-1} (PP-g-MA). A strong anhydride peak confirms coupling occurred.
- Regulatory Note: Confirm that your rLDPE feedstock is certified for the intended application (e.g., non-food contact). Even with perfect additives, contaminated recycled resin cannot be made safe.

Remarks

This formulation turns a low-value, problematic waste stream into a high-volume, economically viable material.

It doesn't fix degradation; it works *around* it.

Use this as your go-to for circular economy compounding.

Test it under real recycling line conditions.

And never underestimate the power of a well-placed coupling agent, even in broken polymer chains.

15.7 Silicone-Based Wetting Additive for Film Extrusion

This formulation is designed for low-density polyethylene (LDPE) films used in high-speed packaging applications, such as food wraps, shrink films, and laminated pouches, where surface smoothness, low coefficient of friction, and high gloss are essential for downstream processing (printing, sealing, filling) and end-user appeal.

In film extrusion, surface defects, haze, stickiness, or poor sealing, are often not caused by poor melt homogeneity, but by inadequate surface wetting and lubrication. Standard slip agents like erucamide alone cannot prevent surface tackiness under high line speeds or elevated temperatures. Adding a PDMS-based silicone masterbatch enhances surface mobility, reduces die build-up, and promotes uniform melt flow at the die exit, critical for achieving Class A surface quality.

This system combines PDMS for surface wetting, erucamide for slip, and Irganox 1010 to prevent oxidative degradation during prolonged extrusion. The result is a film that runs cleanly on high-speed lines, seals reliably, and presents a glossy, non-stick surface, without blooming or migration issues.

"You can't polish a film after it's extruded. If the surface doesn't look perfect as it leaves the die, you've already lost the batch."
Film Extrusion Supervisor, Flexible Packaging, Brazil

15.7.1 Composition and Functionality

This LDPE film formulation isn't about chasing gloss, it's about delivering a surface that runs cleanly, seals reliably, and stays non-sticky under high-speed winding and printing. It uses a 3.0% PDMS silicone masterbatch to create a uniform, low-energy surface layer, paired with 1.5% erucamide for controlled slip and 0.5% Irganox 1010 to prevent thermal degradation during long extrusion runs. No waxes, no EBS, those cause haze or interfere with print adhesion. Every ingredient here serves a single purpose: keep the film surface smooth, consistent, and free from die build-up, hour after hour. If your film sticks, smears, or rejects in printing, this is the functional baseline that keeps you running.

Table 15.7.1: Silicone-Based LDPE Film System: Composition & Functionality

Ingredient	CAS #	Percentage (%)	Function
LDPE	9002-88-4	95.0	Base Polymer
Polydimethylsiloxane (PDMS)	63148-62-9	3.0	Surface Wetting Agent
Antioxidant (Irganox 1010)	6683-19-8	0.5	Thermal Stabilizer
Slip Agent (Erucamide)	112-84-5	1.5	Surface Modifier

Why This Blend Works:

- PDMS is not a dispersant, it's a *surface modifier*. It does not interact with the bulk polymer but forms a low-energy, lubricious layer at the air-polymer interface.
- Erucamide provides long-term slip, but alone causes haze and stickiness at high speeds. PDMS suppresses this by preventing erucamide from forming thick, uneven layers.
- Irganox 1010 is essential, PDMS degrades under prolonged heat, and LDPE oxidizes easily during long extrusion runs. The antioxidant extends processing window and prevents yellowing.
- No other dispersants (e.g., PE wax, EBS) are included, they interfere with surface migration and cause bloom or printability issues.

15.7.2 Processing Recommendations

Film extrusion with this LDPE/silicone system demands control, not speed. The goal is to let PDMS and erucamide migrate fully to the surface before the melt solidifies, too fast a quench, and you trap them in the bulk, causing haze and poor slip. Stick to a narrow 190–205°C melt window; above

210°C, PDMS volatilizes and creates pinholes, odor, and plate-out. Keep screw speed moderate (30–50 RPM) to avoid melt fracture, and always use mirror-polished dies, rough surfaces defeat the purpose of a surface-active additive. Chill rolls at 15–20°C lock in the migrated layer without thermal shock. And don't feed additives separately; pre-blend everything so the system works as a unit. This isn't about pushing line speed, it's about engineering the surface, one micron at a time.

- Extruder Type: Single-screw or twin-screw co-rotating extruder ($L/D \geq 25$) with film die (annular or flat die).
- Temperature Profile:
  - Feed Zone: 160–170°C
  - Melting Zone: 180–190°C
  - Metering Zone: 190–200°C
  - Die Zone: 195–205°C (Do not exceed 210°C)

Critical: PDMS begins to volatilize above 210°C, causing odor, fumes, and surface pinholes.

- Screw Speed: 30–50 RPM, high speed causes melt fracture. Use lower speed with high die pressure for smooth film.
- Residence Time: Minimum 3 minutes, allows PDMS and erucamide to migrate to surface before quenching.
- Die Design: Use mirror-polished die land ($Ra < 0.2$ µm). PDMS works best on smooth surfaces.
- Quenching: Use chilled water rolls at 15–20°C to rapidly solidify surface and lock in PDMS layer.
- Dosing: Pre-blend all additives with LDPE pellets in a high-shear mixer for 5 minutes. Feed via main hopper.

Do not feed PDMS masterbatch via side feeder, it requires full melt exposure to migrate properly.

Pro Tip: Run a "bloom test" after 24 hours: rub film surface with clean cloth. If residue transfers, erucamide is overdosed. Reduce to 1.0%.

15.7.3 Key Performance Indicators (KPIs)

In LDPE film with silicone and erucamide, performance isn't judged by how it extrudes, it's proven by what you can measure on the line and in the roll. This section focuses on the six KPIs that really matter: surface gloss ≥85 GU (critical for premium packaging), coefficient of friction ≤0.3 (to keep high-speed lines running), haze ≤2.0% (for clarity in food wrap), anti-block ≤2% after 24h at 40°C, seal strength ≥1.5 N/mm (to avoid leaks), and die build-up frequency >72 hours. These aren't lab ideals; they're the same thresholds packaging converters use to accept or reject a batch. If your film misses even one, it's not "close enough." It's scrap, rework, or a customer complaint waiting to happen. Measure it, control it, ship it.

Table 15.7.3: LDPE Film KPIs

KPI	Target	Measurement Method
Surface Gloss (60°)	≥ 85 GU	ASTM D523 (film sample, 25μm thickness)
Coefficient of Friction (COF)	Static ≤ 0.3, Dynamic ≤ 0.2	ASTM D1894 (film-to-film)
Anti-Block Performance	≤ 2% blockage after 24h @ 40°C	ASTM D3452 (stacked film under load)
Haze (%)	≤ 2.0%	ASTM D1003
Seal Strength (ASTM F88)	≥ 1.5 N/mm	Heat seal test at 130°C, 0.5s dwell
Die Build-Up Frequency	> 72 hours between cleaning	Visual inspection of die lip
Printability (Adhesion)	> 95% ink adhesion after 24h	ASTM D3359 (cross-hatch tape test)

A COF > 0.4 or haze > 3% = failure. PDMS must migrate fully before quenching to be effective.

15.7.4 Compatibility & Synergy

In LDPE film formulations, surface performance lives or dies by what you mix, and what you don't. This section cuts through trial-and-error with a field-tested compatibility map: how PDMS and erucamide work together to deliver slip and gloss without haze, and why adding even small amounts of PE wax, calcium stearate, or EBS sabotages optical clarity and printability. You'll also see where synergy pays off, like pairing PDMS with high-purity TiO_2 for brighter whites, and where it backfires, like using silicone in films destined for lamination or metallization. No theory, just the additive pairings that keep your film running cleanly, sealing reliably, and looking sharp off the roll.

✅ Synergistic With:

- Erucamide, PDMS prevents erucamide from forming thick, sticky layers, enabling lower dosage and better slip.
- Irganox 1010, Stabilizes both LDPE and PDMS, extending extrusion run time.
- Colorants (TiO_2, organic pigments), PDMS improves pigment wetting at surface, enhancing gloss.
- Metallized films, reduces surface tack, preventing metal layer adhesion to roll.

❌ Avoid Combining With:

- Waxes (PE, PP, Montan), Compete with PDMS for surface migration; cause haze and reduce gloss.
- Silicone oils (non-masterbatch), Volatilize, cause fumes, and leave oily residues.

- Talc or $CaCO_3$ fillers, increase surface roughness; negate PDMS's gloss benefits.
- High-MFI LDPE (MFI > 5), Too fluid; PDMS migrates too fast, causing bloom and print adhesion failure.

Pro Tip: PDMS works best in low-filler, low-additive systems. If you add fillers, expect a 10–15% drop in gloss.

15.7.5 Advantages / Highlights

This LDPE film formulation delivers the surface performance demanded in high-speed packaging, consistent gloss, reliable slip, and clean sealing, without the haze, plate-out, or printability issues that plague generic additive packages. The 3.0% PDMS masterbatch creates a uniform, low-energy surface layer that stays intact under winding, printing, and heat sealing, while 1.5% erucamide provides controlled slip without excessive bloom. Irganox 1010 stabilizes the system during long extrusion runs, preventing the thermal degradation that causes odor and specking. No waxes, no EBS, no compromises: just a clean-running, FDA-compliant system that keeps film lines moving and converters out of rework.

- Exceptional gloss, achieves near-mirror finish without post-treatment.
- Low friction, Enables high-speed packaging line speeds (>200 m/min).
- Superior anti-block, eliminates film sticking during storage and winding.
- Clean processing, reduces die build-up, extending run time and reducing downtime.
- Print-ready surface, Excellent ink adhesion without corona treatment.
- Thermally stable, Irganox 1010 prevents degradation during long extrusion runs.

15.7.6 Typical Applications

This LDPE film formulation is built for high-speed packaging lines where surface performance can't be an afterthought, it has to work from the moment the melt hits the die. It's commonly used in food wraps, shopping bags, stretch films, and laminated pouches that demand consistent slip, high gloss, and reliable heat sealing without plate-out or printability issues. The combination of PDMS and erucamide ensures the film runs cleanly through high-output lines, stays non-sticky during winding and storage, and provides a smooth, printable surface straight off the roll. If your film is meant to move fast, look sharp, and seal every time, without constant die cleaning or customer complaints, this is the functional baseline that delivers in real-world production.

- Food Packaging Films, Sandwich wraps, cheese pouches, frozen food bags
- Shrink Films, Beverage multi-packs, clamshells
- Laminated Films, Barrier films for snacks, coffee, pet food
- Stretch Films, Pallet wrap (for low-tack surface)
- Medical Films, Sterile packaging (non-toxic, low-outgassing)

Ideal for applications requiring high-speed processing, visual appeal, and consistent sealing performance.

15.7.7 Handling & Safety Notes

Handle this LDPE/silicone film formulation like any industrial compound, focus on basics, not hype. PDMS and erucamide are low-hazard, but fine powders generate dust; wear an N95 mask and use local exhaust ventilation during dry blending. Avoid prolonged skin contact with additives, and always use safety glasses and gloves. LDPE is flammable, keep material away from open flames and hot surfaces. Most importantly, never skip drying: moisture in the resin causes voids, surface bubbles, and inconsistent slip performance. No exotic risks here, but ignore fundamentals and you'll pay in scrap, downtime, or safety incidents. Good handling isn't paperwork; it's keeping your team and your line running without surprises.

- Dust Control: LDPE pellets and erucamide generate fine dust, use N95 masks and local exhaust ventilation.
- Skin Contact: All components are non-irritating. Avoid prolonged contact with dry powders.
- Fire Hazard: LDPE is combustible. Store away from ignition sources.
- PPE Required: Gloves, safety glasses, and respirator during dry blending.
- No known acute toxicity, SDS compliant with GHS standards.
- PDMS masterbatch is non-toxic and REACH-compliant for food contact when used within dosage limits.

15.7.8 Packaging and Storage

Store this LDPE/silicone film formulation dry, sealed, and cool, any exposure to moisture or heat invites haze, plate-out, or inconsistent slip performance. PDMS and erucamide are stable when protected, but LDPE absorbs moisture, and even small amounts cause voids or surface defects during high-speed extrusion. Keep bags in a climate-controlled space (15–25°C, <40% RH), use strict FIFO inventory, and always pre-dry before processing.

Packaging is typically 25 kg multi-wall paper bags with a PE liner for the compound, 20 kg HDPE pails for PDMS masterbatch and erucamide, and 1 kg foil pouches for Irganox 1010. Avoid stacking more than three bags high, compaction can trap moisture and damage seals. Shelf life is 18 months for LDPE, 18 months for PDMS masterbatch, and 24 months for other additives, if stored dry and dark.

Before use, inspect for clumping, off-color, or oily surface residue. If in doubt, dry at 70°C for 4–6 hours and run a small validation batch. This isn't just raw material, it's a surface-critical system. Treat it like one.

- Packaging:
    - LDPE pellets: 25 kg multi-wall paper bags with PE liner.

- PDMS masterbatch: 20 kg HDPE pails (sealed).
- Erucamide: 20 kg HDPE drums.
- Irganox 1010: 1 kg foil pouches or 25 kg bags.

- Storage Conditions:
  - Temperature: 15–25°C
  - Humidity: <40% RH, moisture causes LDPE to hydrolyze slightly, reducing melt strength.
- Shelf Life:
  - LDPE: 12 months
  - PDMS Masterbatch: 18 months
  - Erucamide: 24 months
  - Irganox 1010: 24 months
- Handling: Use FIFO inventory. Store PDMS masterbatch away from direct sunlight, UV degrades silicone over time.

15.7.9 Optimization Tips / Remarks

This LDPE film formulation works reliably as-is, but it rewards careful, hands-on tuning. If haze creeps in during high-speed runs, it's often not the pigment, it's erucamide migrating unevenly; try reducing it to 1.0% and relying more on PDMS for surface control. For deep reds or blues with high-tinting organic pigments, pre-disperse the colorant with stearic acid before compounding, it prevents specking without adding cost. Always verify pigment dryness: LDPE is hygroscopic, and even 0.03% moisture causes micro-voids that scatter light and kill gloss. And skip silicone additives entirely if the film will be printed or laminated, they pass visual inspection but fail adhesion tests down the line. This isn't about fixing what's broken, it's about tightening a solid baseline so it holds up under real-world processing, shift after shift.

Optimization Tips

- For Higher Gloss (>90 GU): Increase PDMS to 3.5–4.0%. Monitor COF, if it drops below 0.15, the film becomes too slippery for winding.
- For High-Speed Lines (>250 m/min): Reduce erucamide to 1.0% and increase PDMS to 3.5%, faster surface migration is critical.
- Troubleshooting Haze or Bloom:
    - If film appears hazy → Check for wax contamination in LDPE or extruder.
    - If film feels oily → Reduce PDMS to 2.5%, over-migration causes surface residue.
- For Improved Seal Strength: Increase Irganox 1010 to 0.7%, oxidation at seal zone reduces bonding.
- Upgrade Path: For metallized films, replace erucamide with oleamide, lower melting point, better migration at low temperatures.
- Quality Control Tip: Run a FTIR scan of the film surface. Look for the Si–O–Si peak at 1000–1100 cm^{-1}. A strong, sharp peak confirms PDMS migration.
- Regulatory Note: Confirm your PDMS masterbatch is certified for FDA 21 CFR 178.2010 and EU 10/2011 for direct food contact. Request a Letter of Guarantee (LOG).

Remarks

- This formulation is not a "magic bullet" for film gloss, it's a precision-engineered surface control system.
- It doesn't fix bad processing.
- It enables *perfect* processing.
- Use this as your standard for premium film applications.
- Test it under real line speeds.

- And never compromise on surface chemistry, because in film extrusion, the surface is the product.

15.8 High-Temperature Dispersant System for PBT Compounds

This formulation is designed for glass fiber-reinforced PBT compounds used in high-temperature electrical connectors, motor housings, relay bases, and under-hood automotive components, applications requiring dimensional stability, high heat deflection temperature (HDT), and consistent mechanical performance under prolonged exposure to 150–180°C.

PBT is a semi-crystalline engineering resin with excellent thermal and chemical resistance, but it is notoriously difficult to process with glass fibers. Fibers tend to clump, break excessively during compounding, and debond from the matrix due to poor interfacial wetting. Standard dispersants like EBS or PE wax degrade above 240°C, leading to smoke, odor, and mechanical failure.

This system employs a phosphate ester wetting agent for high-temperature compatibility and superior fiber wetting, paired with a silicone wax that provides lubrication without migration or degradation. Together, they enable uniform fiber distribution, reduced melt viscosity, and minimal fiber breakage, even at 260–270°C processing temperatures.

"In PBT, you don't just disperse glass fiber, you protect it. If the fiber breaks or pulls out, your part fails electrically. This system keeps the fiber intact and bonded."
Electrical Components Engineer, Automotive Tier-1, USA

15.8.1 Composition and Functionality

This PBT formulation is built for one thing: keeping glass fibers intact and bonded under high heat and electrical stress. It uses a phosphate ester wetting agent to actively bond to fiber surfaces, something EBS or PE wax can't do, and a high-molecular-weight silicone wax to lubricate the melt without migrating, degrading, or fouling electrical contacts. No low-MW lubricants, no metal soaps, no amine-based additives that break down above 240°C and leave voids or odor. The result is a clean-running, thermally stable compound that delivers consistent fiber distribution, high HDT (≥230°C), and reliable electrical insulation, exactly what's needed for connectors, relay bases, and under-hood components that can't afford field failure. If your fibers aren't wetted, they're not working. Here, they're locked in.

Table 15.8.1: High-Temp PBT System: Composition & Functionality

Ingredient	CAS #	Percentage (%)	Function
PBT Resin	24968-12-5	92.0	Base Polymer
Phosphate Ester Wetting Agent	N/A	1.5	Wetting / Flow Aid
Silicone Wax	N/A	2.0	Dispersant / Lubricant
Glass Fiber	65997-17-3	4.5	Reinforcement

Why This Blend Works:

- Phosphate ester is chemically stable at PBT processing temperatures and actively wets glass fiber surfaces, unlike fatty acid derivatives that decompose.
- Silicone wax (high-MW, non-volatile) lubricates the melt without blooming or interfering with electrical properties.
- No EBS, no PE wax, no metal soaps, all degrade or volatilize above 240°C, causing defects and odors.
- The low dosage (3.5% total additive) minimizes cost and avoids plasticization of the PBT matrix.

15.8.2 Processing Recommendations

Processing this PBT compound isn't about brute force, it's about protecting the glass fibers while ensuring the phosphate ester and silicone wax fully wet every surface. Run the extruder just hot enough to melt the resin (260–270°C max) but not so hot that the additives degrade or the PBT chains hydrolyze. Feed glass fiber downstream, never with the resin, so it's coated immediately, not pounded. Keep residence time at 3.5 minutes minimum to let the phosphate ester bond to fiber surfaces, and always dry the resin to <0.02% moisture or you'll see voids, weak spots, and inconsistent flow. This isn't a forgiving system; skip one step, and you'll pay in fiber breakage, electrical failures, or die buildup. Precision here isn't optional, it's what keeps the part from failing in the field.

- Extruder Type: Twin-screw co-rotating extruder (L/D ≥ 40) with side feeder for glass fiber.
- Temperature Profile:
  - Feed Zone: 230–240°C (keep cool to prevent premature fiber wetting and degradation)
  - Melting Zone: 250–260°C (ensure complete melting of PBT crystals)
  - Metering Zone: 260–270°C (Do not exceed 275°C)

Critical: PBT degrades above 275°C. Use inline melt sensors to verify actual temperature.

- Screw Speed: 100–120 RPM, high enough to disperse fibers, but low enough to minimize breakage.
- Residence Time: Minimum 3.5 minutes, critical for phosphate ester to fully wet fiber surfaces.
- Venting: Apply vacuum at Zone 2 to remove moisture and any trace volatiles. Dry PBT to <0.02% moisture before compounding.
- Fiber Feeding: Feed glass fiber via downstream side feeder at Zone 3, never in the main hopper. Pre-treat fibers with silane coupling agent for best results.
- Pelletizing: Use underwater pelletizer with chilled water (15–20°C) to rapidly solidify melt and lock in fiber orientation.

- Dosing: Pre-blend phosphate ester and silicone wax with PBT pellets. Feed glass fiber separately via side feeder.

Pro Tip: If fiber breakage is high, reduce screw speed by 10% and increase kneading block spacing in Zones 4–5 to reduce shear.

15.8.3 Key Performance Indicators (KPIs)

In high-temperature PBT compounds with glass fiber, performance isn't judged by how smoothly it extrudes, it's proven by numbers that matter on the production floor and in the field. This section focuses on the seven KPIs that separate reliable formulations from rejects: tensile strength ≥120 MPa, notched impact ≥8 kJ/m², HDT ≥230°C, fiber length retention >85%, MFI 18–24 g/10min, die build-up frequency >48 hours, and electrical resistivity >10^{14} Ω·cm. These aren't lab curiosities; they're the same thresholds used by automotive and electrical OEMs to approve or scrap a batch. If your compound misses even one, it's not "close enough." It's a rework ticket or a field failure waiting to happen. Measure it, control it, ship it.

Table 15.8.3: PBT Compound KPIs

KPI	Target	Measurement Method
Tensile Strength (ASTM D638)	≥ 120 MPa	Injection-molded bars
Notched Impact Strength (ASTM D256)	≥ 8 kJ/m²	Charpy impact at 23°C
Heat Deflection Temp (HDT @ 1.82 MPa)	≥ 230°C	ASTM D648
Fiber Length Retention (SEM)	> 85% of original length	Microscopy of fractured surface
Melt Flow Index (MFI)	18–24 g/10min (260°C, 5kg)	ASTM D1238

Die Build-Up Frequency	> 48 hours between cleaning	Visual inspection of extrusion die
Electrical Resistivity (ASTM D257)	> 10^{14} Ω·cm	For insulation applications
Batch-to-Batch Consistency	±3% variation in tensile strength	SPC control chart

A drop in HDT below 220°C or fiber length retention below 80% = poor wetting or excessive shear. Revisit additive dosage or screw design.

15.8.4 Compatibility & Synergy

In PBT compounds, getting glass fiber to stay put isn't just about the fiber, it's about what's in the mix with it. This section cuts through trial-and-error with a field-tested compatibility map: how phosphate ester bonds chemically to both PBT's ester groups and glass fiber's silanol surfaces, while high-MW silicone wax lubricates cleanly without migrating or compromising electrical properties. It also shows where things quietly fail, like EBS or erucamide degrading above 240°C and releasing amines that hurt mechanicals, or zinc stearate catalyzing PBT breakdown. No theory, just the additive pairings that keep your connectors dimensionally stable, electrically reliable, and running cleanly for 48+ hours. If your additives are working against each other, you're not compounding, you're compromising.

✅ Synergistic With:

- Glass Fiber (E-glass), Phosphate ester bonds to silanol groups on fiber surface, reducing breakage and improving load transfer.
- Flame Retardants (e.g., brominated or phosphorus-based), Silicone wax improves flow without interfering with FR dispersion.
- Nucleating Agents (e.g., sorbitol), Enhances crystallization rate and HDT without affecting dispersion.

- Antioxidants (Irganox 1076), Stabilizes PBT against oxidative degradation during high-temp processing.

✕ Avoid Combining With:

- EBS or Erucamide, decompose above 240°C, releasing amines, causing odor, discoloration, and reduced HDT.
- Low-MW Silicone Oils, Volatilize, cause voids, and contaminate electrical contacts.
- Metal Soaps (e.g., CaSt, ZnSt), Catalyze PBT degradation and reduce electrical insulation properties.
- High-MFI PBT (MFI > 30), Too fluid; leads to poor fiber orientation and warpage.

Pro Tip: Always use silane-treated glass fiber with this system. The phosphate ester + silane combination creates a dual-bonding mechanism for maximum adhesion.

15.8.5 Advantages / Highlights

This PBT formulation delivers the thermal stability, fiber integrity, and surface cleanliness required for high-reliability electrical and automotive parts, without the smoke, odor, or plate-out that come from standard lubricants like EBS or PE wax. The combination of phosphate ester and high-MW silicone wax ensures glass fibers stay fully wetted and intact through compounding, preserving tensile strength and impact resistance even after repeated thermal cycling. Melt flow stays consistent for 48+ hours, screen pressure remains stable, and parts come off the mold with a smooth, print-ready surface. No migratory additives, no volatile breakdown products, just a lean, high-temperature system that performs under pressure and meets stringent electrical and flame-retardant specs.

- Exceptional fiber retention, minimizes breakage, preserving mechanical strength.

- High thermal stability, all components survive 270°C processing without degradation.
- Low die build-up, Silicone wax prevents resin adhesion, enabling long production runs.
- Excellent electrical properties, No conductive residues or ionizable byproducts.
- Smooth surface finish, No specks, streaks, or roughness on molded parts.
- Cost-effective, Low additive load (3.5%) keeps cost down while delivering premium performance.

15.8.6 Typical Applications

This PBT compound is built for parts that must hold up under high heat, electrical load, and mechanical stress without cracking, warping, or failing insulation, conditions common in under-hood automotive components, electrical connectors, relay bases, and industrial motor housings. The combination of phosphate ester and high–molecular-weight silicone wax ensures glass fibers stay intact and fully wetted, delivering consistent mechanicals, low die build-up, and a clean, printable surface straight off the mold. If your application can't tolerate field failures, specking, or loss of dielectric strength, even after thermal cycling or long-term storage, this is the formulation that meets the bar, batch after batch.

- Electrical Connectors & Sockets, High-voltage insulation, creepage resistance
- Motor Housings & End Caps, Under-hood automotive, industrial motors
- Relay Bases & Switch Components, Thermal cycling resistance
- Industrial Gear Wheels, High-stiffness, low-friction applications
- Appliance Housings, Washing machine motors, dryer components

Ideal for parts requiring long-term performance under thermal stress and electrical load.

15.8.7 Handling & Safety Notes

Handle this PBT/glass fiber compound like any industrial material, focus on the basics, not the drama. PBT resin and glass fiber generate fine dust during blending; wear an N95 mask and use local exhaust ventilation to protect lungs and eyes. Phosphate ester and silicone wax are low-hazard, but dry powders can still irritate skin and airways, safety glasses and gloves are non-negotiable. Keep material away from open flames: PBT is self-extinguishing but can release CO and CO_2 when burned. Always dry PBT to <0.02% moisture before processing, wet resin hydrolyzes in the barrel, causing voids, fiber dewetting, and mechanical failure. No exotic risks here, but skip fundamentals and you'll pay in scrap, downtime, or worse. Good handling isn't paperwork. It's keeping your team and your line running without surprises.

- Dust Control: PBT resin and glass fiber generate fine particulates, use N95 masks and local exhaust ventilation.
- Skin Contact: Phosphate ester and silicone wax are non-irritating. Avoid prolonged contact with dry powders.
- Fire Hazard: PBT is self-extinguishing but can release CO and CO_2 when burned. Store away from ignition sources.
- PPE Required: Gloves, safety glasses, and respirator during dry blending.
- No known acute toxicity, SDS compliant with GHS standards.
- Phosphate ester is REACH-compliant and RoHS-free.

15.8.8 Packaging and Storage

Store this PBT/glass fiber compound dry, cool, and sealed, any moisture or heat exposure will degrade performance before it even hits the extruder. PBT is hygroscopic, and phosphate ester loses effectiveness if it absorbs water; both lead to poor fiber wetting, voids, and inconsistent mechanicals. Keep bags in a climate-controlled space (15–25°C, <30% RH), use strict FIFO inventory, and never skip pre-drying before processing.

Packaging is typically 25 kg multi-wall paper bags with a PE liner for PBT resin, 20 kg HDPE pails for phosphate ester and silicone wax, and sealed moisture-barrier bags for glass fiber. Avoid stacking more than three bags high, glass fiber bags can rupture, and PBT pellets trap moisture when compacted. Shelf life is 12 months for PBT (if vacuum-sealed), 18 months for phosphate ester, and 24 months for silicone wax, if stored dry and dark.

Before use, inspect for clumping, discoloration, or dusty surfaces. If in doubt, dry PBT at 120°C for 4–6 hours and run a small validation batch. This isn't just raw material, it's a precision-engineered system for high-reliability parts. Treat it like one.

- Packaging:
  - PBT resin: 25 kg multi-wall paper bags with PE liner (vacuum-sealed).
  - Phosphate ester: 20 kg HDPE drums (light-sensitive; store in dark).
  - Silicone wax: 20 kg HDPE pails.
  - Glass fiber: 25 kg sealed bags (moisture-sensitive).
- Storage Conditions:
  - Temperature: 15–25°C
  - Humidity: <30% RH, PBT and glass fiber are hygroscopic.
- Shelf Life:
  - PBT: 12 months (if vacuum-sealed)
  - Phosphate Ester: 18 months
  - Silicone Wax: 24 months
  - Glass Fiber: 12 months (if sealed)
- Handling: Use FIFO inventory. Always dry PBT at 120°C for 4–6 hours before compounding.

15.8.9 Optimization Tips / Remarks

This PBT formulation works reliably as-is, but it rewards smart, hands-on tuning. If fiber length retention drops below 85%, it's often not the additive, it's shear. Reduce screw speed by 10% and add reverse kneading elements to protect fibers during compounding. For electrical connectors that fail leakage tests, verify that no EBS or zinc stearate is present, even trace amounts from previous runs can create conductive pathways. If die build-up appears after 24 hours, increase silicone wax to 2.2% but never exceed 2.5%, or surface gloss will suffer. And always dry PBT to <0.02% moisture: wet resin hydrolyzes in the barrel, causing voids that show up only under high-voltage testing. This isn't about chasing specs, it's about locking in performance that survives real-world thermal cycling, electrical load, and production variability, shift after shift.

Optimization Tips

- For Higher Fiber Loadings (6–8%): Increase phosphate ester to 2.0% and silicone wax to 2.5%. Monitor MFI, if it drops below 15 g/10min, reduce screw speed, not temperature.
- For Improved Impact Strength: Add 1–2% PBT-g-MA (maleated PBT), enhances fiber-matrix bonding further.
- Troubleshooting Fiber Breakage:
    - If fiber length is <80% → Reduce screw speed by 10–15% and use longer kneading elements.
    - If fiber clumps → Check for moisture in fiber or resin. Dry longer.
- For Electrical Applications: Avoid any additives containing sulfur or chlorine, they can corrode contacts.
- Upgrade Path: For ultra-high-temperature applications (>180°C continuous), replace silicone wax with PTFE-based lubricant masterbatch, lower friction, higher thermal stability.

- Quality Control Tip: Run a TGA (Thermogravimetric Analysis) on your compound. Degradation onset should be >280°C. A lower onset indicates contamination or degradation.
- Regulatory Note: Confirm phosphate ester is UL 94 V-0 certified for flame retardant systems and REACH compliant for global markets. Request a Letter of Guarantee (LOG).

Remarks

- This formulation is not a generic "PBT additive mix."
- It is a high-temperature, precision-engineered system built for the unforgiving demands of electrical and automotive engineering.
- Use it as your baseline for high-performance PBT.
- Test it under real thermal and shear stress.
- And never compromise on fiber wetting, because in PBT, the fiber is the backbone, and the dispersant is its lifeline.

15.9 Universal Dispersing Additive for Color Masterbatches

This formulation is designed as a universal carrier system for color masterbatches intended for use across multiple non-polar and moderately polar thermoplastics, including PE, PP, PS, and some EVA-based compounds. Its purpose is to offer a single, cost-effective masterbatch base that delivers adequate pigment wetting, good flow, and acceptable dispersion across diverse applications, eliminating the need for multiple carrier resins and reducing inventory complexity.

While not a high-performance solution for demanding applications (e.g., automotive, medical, or high-gloss), this system provides a reliable, "good enough" option for general-purpose packaging, consumer goods, and low-cost injection-molded parts where visual consistency matters more than mechanical precision. It combines EVA as a carrier (offering broader compatibility than PE or PP alone), PE wax for wetting, and EBS for lubrication and flow, a classic, widely used triad in the masterbatch industry.

"You don't need a Ferrari to haul bricks. This is the pickup truck of masterbatch carriers, simple, tough, and gets the job done across 80% of our lines."
Masterbatch Production Manager, Packaging Converter, India

15.9.1 Composition and Functionality

This universal masterbatch system isn't built for perfection, it's built for practicality. It uses EVA as a carrier because it plays reasonably well with PE, PP, and PS, avoiding the need for multiple dedicated carriers in shops running mixed-resin jobs. PE wax and EBS deliver baseline wetting and flow at a low cost, just enough to keep pigment agglomerates under control in non-critical applications like shopping bags, toys, or industrial bins. There's no high-end chemistry here, no maleated polymers, no specialty dispersants, because the goal isn't to win a gloss contest, but to cut inventory, simplify changeovers, and get color out the door without rework. If your priority is speed, simplicity, and cost over premium performance, this is the workhorse that keeps the line moving.

Table 15.9.1: Universal Masterbatch: Composition & Functionality

Ingredient	CAS #	Percentage (%)	Function
EVA Copolymer	24937-78-8	93.0	Carrier Resin
PE Wax	9002-88-4	2.0	Dispersant
EBS	110-30-5	1.0	Wetting Agent
Pigment Mixture	N/A	4.0	Colorant

Why This Blend Works:

- EVA is the key innovation: its polar vinyl acetate groups provide *just enough* polarity to wet a wider range of pigments than pure PE or PP, while remaining compatible with non-polar resins.
- PE wax provides rapid, low-cost wetting, ideal for the initial dispersion phase.

- EBS is a workhorse: it improves flow, reduces die build-up, and enhances surface feel.
- The low pigment load (4%) ensures easy dispersion and avoids overloading the system.
- This is a cost-optimized compromise, not a premium solution, but one that eliminates the need for 3–4 different carrier systems.

15.9.2 Processing Recommendations

This universal masterbatch system runs on standard equipment, but don't treat it like a high-performance formulation. Keep melt temperatures at or below 200°C; EVA degrades above 205°C, releasing acetic acid that causes odor, yellowing, and plate-out. Use a twin-screw extruder if you can, but a single-screw will work for low-shear, non-critical runs, just expect slightly higher speck counts. Pre-blend pigment with PE wax and EBS before adding to EVA to ensure even wetting, and avoid dry pigment feeding, it leads to bridging and inconsistent color. Residence time should be at least 2.5 minutes; less than that and the additives won't fully wet fine pigments like phthalocyanines. This isn't a system that forgives rushed cycles or sloppy blending, it's a cost-effective compromise that works reliably only when you respect its limits.

- Extruder Type: Twin-screw extruder (L/D ≥ 30) recommended. Single-screw acceptable for low-volume, non-critical lines.
- Temperature Profile:
  - Feed Zone: 160–170°C (keep cool to prevent premature melting of EVA)
  - Melting Zone: 180–190°C
  - Metering Zone: 190–200°C (Do not exceed 205°C)

Critical: EVA degrades above 205°C, releasing acetic acid and causing odor and discoloration.

- Screw Speed: 80–100 RPM, sufficient for pigment deagglomeration without excessive shear.
- Residence Time: Minimum 2.5 minutes, ensures full wetting of pigments by PE wax and EBS.
- Venting: Apply vacuum at Zone 2 if using moisture-sensitive pigments (e.g., carbon black).
- Pelletizing: Use air-cooled strand pelletizer. Avoid water quenching unless part is for low-temp applications.
- Dosing: Pre-blend pigment with PE wax and EBS in a high-shear mixer for 5–7 minutes. Feed this pre-mix into EVA pellets via main hopper.

Pro Tip: This system works best with pre-dispersed pigment concentrates. Avoid dry pigment powders, poor dispersion leads to specks.

15.9.3 Key Performance Indicators (KPIs)

This universal masterbatch system isn't built for perfection, it's built to keep your line running across multiple resins without constant reformulation. That means KPIs focus on what matters in real-world, high-mix environments: visual consistency, acceptable color match, and stable flow, not lab-grade performance. The target is ≤5 specks per 10 cm², ΔE ≤2.0, MFI between 6.0–8.5 g/10min, and filter pressure rise under 10 bar over 2 hours. These thresholds won't pass an automotive audit, but they keep job shops, packaging converters, and small injection molders out of rework. If your batch meets these metrics, it ships. If not, it's not the additive, it's the expectation that needs adjusting.

Table 15.9.3: Universal Masterbatch KPIs

KPI	Target	Measurement Method
Visual Speck Count (10x)	≤ 5 specks per 10 cm²	Visual inspection on injection-molded plaques

Color Strength (ΔE*ab)	≤ 2.0 (vs. standard)	Spectrophotometer (D65 illuminant)
Melt Flow Index (MFI)	6.0–8.5 g/10min (190°C, 2.16kg)	ASTM D1238
Filter Pressure Rise	≤ 10 bar over 2 hours	Extruder screen pack monitoring
Batch-to-Batch Consistency	±0.5 ΔE	SPC control chart
Surface Finish	No bloom, no haze	Visual inspection under daylight

A ΔE > 2.5 or speck count > 8 per 10 cm^2 *= unacceptable for most applications. This is not a high-end system, manage expectations.*

15.9.4 Compatibility & Synergy

In this universal masterbatch system, "compatibility" doesn't mean perfection, it means practical compromise. EVA gives just enough polarity to help pigments wet in PE, PP, and PS, while PE wax and EBS provide baseline flow and dispersion for non-critical applications. But push too far, and the limits show: EBS migrates in PS, causing haze; PE wax clouds in PET; and neither survives high-temperature processing. This section lays out what actually works together, like EVA with iron oxides or low-load organic pigments, and what quietly undermines performance, like using this system with carbon black or in films that need printability. No theory, just the field-tested boundaries of a cost-driven, multi-resin solution. If your goal is broad coverage with modest expectations, this is your map. If you need peak performance, look elsewhere.

Compatibility & Synergy

✅ Synergistic With:

- PE, PP, PS, EVA, and some LDPE, EVA carrier provides broad compatibility.
- Organic and inorganic pigments, PE wax and EBS provide adequate wetting for moderate-load applications.
- Low-cost stabilizers (e.g., Irganox 1010), Compatible at ≤0.2%.

✕ Avoid Combining With:

- High-performance pigments (e.g., phthalocyanine blue at >5%), EVA lacks the anchoring power for fine, high-surface-area pigments.
- Polar resins (PA, PC, PET), EVA is not polar enough to stabilize pigments in these matrices.
- High-temperature applications (>200°C), EVA degrades, releasing acetic acid and causing odor.
- Silicone additives, can cause surface defects (fish-eyes) and interfere with printing.
- Flame retardants or fillers >1%, Overloads the system; causes poor dispersion and increased viscosity.

Pro Tip: Do not use this as a carrier for black masterbatches with carbon black >3%. Use a PE-g-MA or BYK-P 4100 system instead.

15.9.5 Advantages / Highlights

This universal masterbatch system delivers consistent, acceptable color and flow across PE, PP, PS, and EVA without the complexity of maintaining multiple carrier resins. The EVA base offers just enough polarity to handle a broad range of pigments, while PE wax and EBS provide baseline wetting and lubrication for smooth extrusion and mold release. It's not built for high-gloss or high-stress applications, but it's reliable for everyday parts like shopping bags, toys, and industrial bins where cost, speed, and simplicity matter more than premium performance. All components are FDA-compliant for indirect food contact, and the formulation runs cleanly on standard equipment with no special drying or handling required. If your priority is inventory reduction and batch-to-batch reliability for non-critical colors, this is the workhorse that keeps your shop moving.

- Single carrier for multiple resins, reduces inventory, simplifies logistics, lowers storage costs.
- Low cost, EVA, PE wax, and EBS are among the most economical additives available.
- Good flow and mold release, EBS ensures smooth processing on standard injection machines.
- Easy to use, No special equipment or drying required.
- FDA-compliant components, All ingredients listed under 21 CFR 178.3820 for indirect food contact.
- Ideal for small batch runs, Perfect for job shops and custom color converters.

15.9.6 Typical Applications

This universal masterbatch system is built for job shops, packaging converters, and small injection molders who need one reliable color concentrate that works across multiple resins, without the cost and complexity of maintaining separate carriers for PE, PP, PS, and EVA. It's commonly used in low-to-mid-tier applications like shopping bags, trash liners, toys, storage bins, and basic consumer containers where color consistency matters more than premium gloss or mechanical performance. The EVA-based carrier delivers acceptable dispersion and flow in standard processing equipment, making it ideal for short runs, custom colors, and high-mix environments where setup speed and inventory simplicity are key. If your goal is to keep the line moving and the customer satisfied, not win a design award, this is the practical baseline that gets the job done.

- General Packaging, Shopping bags, trash bags, stretch film (opaque colors)
- Consumer Goods, Toys, buckets, containers, kitchenware
- Industrial Parts, Non-structural housings, cable ties, clips
- Low-Cost Injection Molding, where color consistency is more important than mechanical strength
- Job Shop Masterbatches, for customers requiring multiple colors on limited equipment

Ideal for applications where cost, speed, and simplicity outweigh the need for premium performance.

15.9.7 Handling & Safety Notes

Handle this universal masterbatch like any industrial compound, focus on the basics, not the hype. EVA, PE wax, and EBS generate fine dust during blending; wear an N95 mask and use local exhaust ventilation to protect lungs and eyes. All components are low-hazard, but dry powders can still irritate skin and airways, safety glasses and gloves are non-negotiable. Keep material away from open flames: EVA and PE are combustible. Always dry the masterbatch before processing, moisture in EVA can cause acetic acid release, leading to odor, corrosion, and inconsistent color. No exotic risks here, but skip fundamentals and you'll pay in scrap, downtime, or worse. Good handling isn't paperwork; it's keeping your team and your line running without surprises.

- Dust Control: Pigment handling requires N95 masks and local exhaust ventilation.
- Skin Contact: All components are non-irritating. Avoid prolonged contact with dry powders.
- Fire Hazard: EVA and PE are combustible. Store away from ignition sources.
- PPE Required: Gloves, safety glasses, and respirator during dry blending.

- No known acute toxicity, SDS compliant with GHS standards.
- EVA and EBS are REACH-compliant and non-toxic.

15.9.8 Packaging and Storage

Store this universal EVA-based masterbatch dry, cool, and sealed, moisture and heat are its quiet enemies. EVA absorbs water, and even slight humidity causes pigment clumping, inconsistent color, and voids during extrusion. Keep bags in a climate-controlled space (15–25°C, <50% RH), use strict FIFO inventory, and always pre-dry before high-speed processing.

Packaging is typically 25 kg multi-wall paper bags with a PE liner for the compound, 20 kg HDPE pails for PE wax and EBS, and sealed drums for pigment concentrate. Avoid stacking more than three bags high, compaction traps moisture and risks bag rupture. Shelf life is 18 months for EVA, 24 months for additives, and 12 months for pigment concentrate, if stored dry and dark.

Before use, inspect for caking, off-color, or oily surface residue. If in doubt, dry at 70°C for 4–6 hours and run a small validation batch. This isn't a premium system, but it's a workhorse, treat it with basic care, and it'll deliver consistent color without surprises.

- Packaging:
  - EVA pellets: 25 kg multi-wall paper bags with PE liner.
  - PE wax, EBS: 25 kg bags or 20 kg HDPE drums.
  - Pigment concentrate: 10–20 kg sealed HDPE pails.
- Storage Conditions:
  - Temperature: 15–25°C
  - Humidity: <50% RH, moisture causes EVA to hydrolyze slightly, reducing melt strength.
- Shelf Life:

- EVA: 18 months
- PE Wax / EBS: 24 months
- Pigment Concentrate: 12 months (if sealed)

- Handling: Use FIFO inventory. Store pigment concentrate in dry, dark conditions.

15.9.9 Optimization Tips / Remarks

This universal masterbatch system works reliably as-is, but it rewards smart, hands-on tuning. If specking appears with deep reds or blues, it's usually not the EVA, it's dry pigment agglomerates; pre-disperse colorants with stearic acid or switch to pre-wetted concentrates. If haze shows up in thin films, reduce EBS to 0.8%, even small excess causes blooming in PS or LDPE. Always verify pigment dryness: EVA absorbs moisture, and just 0.05% water leads to micro-bubbles that scatter light and kill gloss. And never use this system for high-gloss, high-heat, or electrical applications, it's built for speed and simplicity, not performance under stress. Treat it like a reliable pickup truck: not flashy, but always ready to move the load.

Optimization Tips

- For Higher Pigment Loadings (5–7%): Increase PE wax to 3.0% and EBS to 1.5%. Monitor MFI, if it drops below 5.0 g/10min, reduce screw speed.
- For Improved Gloss: Add 0.3% Montan wax, but test for bloom risk in storage.
- Troubleshooting Specks or Haze:
  - If specks appear → Check pigment drying. Dry at 80°C for 4 hours.
  - If haze appears → Reduce processing temperature below 200°C.
- For Food Contact: Confirm that your pigment is FDA-compliant, even if the carrier is, the colorant may not be. Always request a Letter of Guarantee (LOG).

- Upgrade Path: For higher-end applications, replace EVA with PP-g-MA or BYK-P 4100, but only if you're willing to pay more and manage multiple carriers.
- Quality Control Tip: Run a DSC scan to confirm EVA melting point remains stable. A drop >5°C indicates thermal degradation.
- Industry Reality:

"We use this for 70% of our colors. For the other 30%, the ones that need to look like a luxury product, we switch to a premium carrier. But for the rest? This saves us a huge amount of money a year in inventory and setup time."
Masterbatch Plant Manager, Mexico

Remarks

- This formulation is not a "magic bullet," but it's a smart, pragmatic solution for high-volume, low-margin, multi-color production.
- It doesn't aim for perfection.
- It aims for reliability, simplicity, and cost control.
- Use it where it belongs:
  Not for the flagship product.
  But for the everyday product that keeps the lights on.

15.10 Bio-Based Wetting Additive for PLA Compounds

Polylactic acid (PLA) is a bio-based, compostable thermoplastic derived from renewable resources such as corn starch or sugarcane. While environmentally attractive, PLA suffers from inherent processing challenges: high melt viscosity, poor pigment wetting, low impact strength, and thermal instability during extrusion or injection molding. Conventional petroleum-based dispersants and lubricants are incompatible with the "green" positioning of PLA and often compromise its biodegradability.

This formulation delivers a 100% bio-based, non-toxic wetting and dispersion system using glycerol monostearate (GMS), epoxidized soybean oil (ESBO), and vegetable-derived bio-wax, all derived from plant sources and certified for food contact and compostability. It enables uniform dispersion of pigments and fillers, improves melt flow, enhances toughness, and maintains the environmental credentials of PLA without sacrificing performance.

"You can't call a PLA part 'sustainable' if you're using a petrochemical dispersant that won't degrade with it. This system keeps the whole lifecycle green, from feedstock to landfill."
Sustainable Materials Engineer, Packaging Innovation, Netherlands

15.10.1 Composition and Functionality

This PLA formulation isn't about chasing performance, it's about making compostable plastic actually work on the line. It uses glycerol monostearate (GMS) to wet pigments without blooming, epoxidized soybean oil (ESBO) to stabilize against thermal degradation, and a bio-based wax to lubricate the melt, all derived from renewable sources and certified for food contact and industrial compostability. No petroleum-based waxes, no EBS, no hidden petrochemicals that sabotage end-of-life claims. The result is a clean-running, bio-based system that reduces brittleness, maintains clarity, and stays true to circularity, without the haze, specking, or plate-out that plagues generic additives in PLA. If your goal is a genuinely sustainable compound that performs reliably in film, injection molding, or sheet extrusion, this is the functional baseline that holds up under real processing conditions.

Table 15.10.1: Bio-Based PLA System: Composition & Functionality

Ingredient	CAS #	Percentage (%)	Function
Polylactic Acid (PLA)	26100-51-6	94.0	Base Resin
Glycerol Monostearate (GMS)	31566-31-1	2.0	Wetting Agent
Epoxidized Soybean Oil	8013-07-8	1.0	Plasticizer

Bio-Wax (Vegetable Source)	N/A	3.0	Dispersant

Why This Blend Works:

- GMS is uniquely compatible with PLA's polar ester groups, unlike paraffin or PE wax, which phase-separate and cause haze.
- ESBO is not just a plasticizer; it actively stabilizes PLA against hydrolysis and thermal degradation, extending processing window.
- Bio-wax provides surface lubrication without blooming, critical for food-contact films and injection-molded containers.
- All components are certified for food contact (FDA, EU 10/2011), industrial compostability (EN 13432), and are non-GMO where applicable.
- No synthetic additives, this system is designed for true circularity.

15.10.2 Processing Recommendations

Processing this PLA formulation demands care, not speed. Keep melt temperatures below 200°C, PLA degrades fast above this threshold, releasing lactic acid that causes yellowing, odor, and embrittlement. Use a twin-screw extruder with vacuum venting to remove moisture and volatiles, and always dry PLA to <0.02% moisture before compounding. Residence time should be at least 3 minutes to let glycerol monostearate fully wet pigments and epoxidized soybean oil stabilize the melt. Avoid water quenching; PLA is sensitive to thermal shock, so use air-cooled strand pelletizing or gentle chill rolls. This isn't a system that forgives shortcuts, skip drying or push the temperature, and you'll pay in haze, specks, or brittle parts that fail on the shelf.

- Extruder Type: Twin-screw co-rotating extruder (L/D ≥ 30) with venting. Single-screw acceptable for low-output film or sheet.
- Temperature Profile:

- Feed Zone: 160–170°C (keep cool to prevent premature degradation)
- Melting Zone: 180–190°C
- Metering Zone: 185–195°C (Do not exceed 200°C)

Critical: PLA degrades rapidly above 200°C, releasing lactic acid, causing yellowing and embrittlement. Use inline melt sensors.

- Screw Speed: 60–80 RPM, high shear increases degradation. Use moderate speed with sufficient residence time.
- Residence Time: Minimum 3 minutes, essential for ESBO to stabilize and GMS to fully wet pigments.
- Venting: Mandatory. Apply vacuum at Zone 2 to remove moisture (>0.05%) and lactic acid vapors. Dry PLA to <0.02% moisture before compounding.
- Pelletizing: Use air-cooled strand pelletizer. Avoid water quenching, PLA is sensitive to thermal shock.
- Dosing: Pre-blend all additives with PLA pellets in a high-shear mixer for 5–7 minutes. Feed via main hopper.

Pro Tip: For colored compounds, pre-disperse pigment with GMS and bio-wax in a small batch before adding to PLA, this ensures even distribution and avoids specks.

15.10.3 Key Performance Indicators (KPIs)

In PLA compounds, performance isn't judged by how it looks off the extruder, it's proven by numbers that hold up under real-world handling, processing, and composting. This section focuses on the seven KPIs that separate functional bio-based parts from brittle, inconsistent failures: MFI ≥10 g/10min (to ensure stable film or injection molding), notched impact strength ≥4 kJ/m² (a 2× improvement over untreated PLA), elongation at break ≥8% (reducing cracking during demolding or storage), ΔE ≤1.5 after 7 days at 60°C (for color stability), surface gloss ≥75 GU (critical for packaging appeal), thermal degradation onset >200°C (TGA-confirmed), and >90% biodegradation in 12 weeks per EN 13432. These aren't optional benchmarks, they're the same thresholds composters, food packagers, and brand owners use to reject or approve a batch. If your compound misses even one, it's not "good enough for compostable." It's scrap. Measure it, control it, ship it.

Table 15.10.3: PLA Compound KPIs

KPI	Target	Measurement Method
Melt Flow Index (MFI)	10–14 g/10min (190°C, 2.16kg)	ASTM D1238, indicates improved flow vs. neat PLA (MFI ~8)
Notched Impact Strength (ASTM D256)	≥ 4 kJ/m²	Charpy test at 23°C, vs. <2 kJ/m² for neat PLA
Tensile Elongation at Break	≥ 8%	ASTM D638, vs. 5–7% for neat PLA
Color Strength (ΔE*ab)	≤ 1.5 (vs. standard)	Spectrophotometer (D65), for consistent color in pigmented films
Surface Gloss (60°)	≥ 75 GU	ASTM D523, critical for transparent containers and packaging
Thermal Stability (TGA)	Onset degradation >200°C	TGA analysis, confirms ESBO's stabilizing effect
Compostability (EN 13432)	>90% biodegradation in 12 weeks	Certified lab testing

A drop in elongation below 6% or MFI below 9 g/10min = insufficient wetting or degradation. Recheck drying or temperature profile.

15.10.4 Compatibility & Synergy

In PLA compounding, "bio-based" doesn't automatically mean "compatible." This section cuts through greenwashing to show what actually works: how glycerol monostearate (GMS) bonds to PLA's ester groups while bio-wax lubricates without blooming, and why even small amounts of petroleum-based EBS or silicone sabotage compostability and cause haze. It also highlights where synergy pays off, like pairing epoxidized soybean oil (ESBO) with bio-wax to stabilize the melt and improve flow, and where it backfires, like using PE wax in PLA film (it phase-separates and clouds the surface). No theory, no fluff, just the additive pairings that keep your PLA running cleanly, composting fully, and performing reliably from extruder to end-of-life.

✅ Synergistic With:

- Natural Pigments (e.g., beetroot, chlorophyll, turmeric), GMS and ESBO enhance wetting without blooming.
- Cellulose Nanofibers (CNF), Bio-wax reduces fiber-fiber hydrogen bonding, improving dispersion.
- Starch Blends (PLA/TPS), ESBO improves compatibility between PLA and hydrophilic starch.
- Antioxidants (e.g., Vitamin E), Compatible; enhances long-term stability.

❌ Avoid Combining With:

- Petroleum Waxes (PE, PP), Cause phase separation and reduce compostability.
- EBS or Stearates, often derived from petrochemicals; may not meet compostability certifications.
- Silicone Additives, contaminate compost streams and are not biodegradable.

- High-MFI PLA (MFI >15), Too fluid; reduces mechanical properties and increases warpage.

Pro Tip: For opaque or colored PLA, use mineral-based pigments (e.g., iron oxides) over synthetic organics, they're more stable under composting conditions.

15.10.5 Advantages / Highlights

This bio-based PLA formulation delivers the surface quality, toughness, and processability needed for real-world compostable packaging, without compromising its environmental credentials. Glycerol monostearate (GMS) and bio-wax effectively wet pigments and fillers without phase separation or haze, while epoxidized soybean oil (ESBO) stabilizes the melt and reduces brittleness. All components are derived from renewable sources, meet FDA and EU food-contact requirements, and are certified for industrial compostability under EN 13432. The result is a clean-running, low-odor compound that extrudes smoothly, seals reliably, and breaks down fully, making it suitable for food trays, cups, mulch films, and other applications where sustainability must be more than a label.

- Fully bio-based and compostable, Meets EN 13432 and ASTM D6400 standards.
- Improves toughness, reduces brittleness without sacrificing stiffness.
- Enhances processability, lowers melt viscosity, reduces torque, and prevents die build-up.
- No bloom or haze, Bio-wax and GMS migrate cleanly, preserving clarity and gloss.
- Food-contact compliant, Safe for packaging, cutlery, and food containers.
- Market differentiator, Enables "100% bio-based, compostable" claims on packaging.

15.10.6 Typical Applications

This bio-based PLA formulation is built for real-world compostable products where appearance, toughness, and processability can't be sacrificed for sustainability. It's commonly used in clear food containers, compostable cutlery, agricultural mulch films, and medical packaging, any application where clarity, seal integrity, and brittleness are make-or-break. The blend of glycerol monostearate, bio-wax, and epoxidized soybean oil ensures pigments stay wetted and the melt flows cleanly on standard extrusion or injection equipment, without the haze, specking, or plate-out that derail most PLA runs. If your goal is a genuinely compostable part that performs reliably on the line and passes EN 13432 certification, this is the practical baseline that holds up under production conditions.

- Food Packaging, Trays, cups, lids, clamshells
- Disposable Cutlery & Straws, Rigid, compostable alternatives to PS/PE
- Agricultural Films, Mulch films that degrade in soil
- Medical Packaging, Sterile trays and blister packs
- Consumer Goods, Toys, cosmetic containers, e-commerce mailers

Ideal for applications where sustainability is a core brand value and performance requirements are moderate.

15.10.7 Handling & Safety Notes

Handle this PLA formulation like any industrial material, focus on the basics, not the buzz. Glycerol monostearate (GMS), epoxidized soybean oil (ESBO), and bio-wax are all low-hazard and derived from natural sources, but they still generate dust and can irritate eyes or lungs during dry handling. Wear an N95 mask, safety glasses, and gloves when blending. PLA itself is combustible, so keep material away from open flames and hot surfaces. Most importantly, always dry PLA to <0.02% moisture before processing, wet resin hydrolyzes in the barrel, causing voids, brittleness, and inconsistent color. No exotic risks here, but skip fundamentals and you'll pay in scrap, downtime, or worse. Good handling isn't paperwork; it's keeping your team and your line running without surprises.

- Dust Control: PLA pellets and bio-wax generate fine dust, use N95 masks and local exhaust ventilation.
- Skin Contact: All components are non-irritating and non-toxic. Safe for handling without gloves (though PPE recommended in industrial settings).
- Fire Hazard: PLA is flammable. Store away from ignition sources.
- PPE Required: Gloves, safety glasses, and respirator during dry blending.
- No known acute toxicity, SDS compliant with GHS standards.
- All components are GRAS (Generally Recognized As Safe) and non-GMO where certified.

15.10.8 Packaging and Storage

Store this PLA formulation dry, cool, and sealed, moisture and heat are its silent enemies. PLA is hygroscopic, and glycerol monostearate (GMS) loses effectiveness when damp; together, they cause clumping, poor pigment wetting, and voids during extrusion. Keep bags in a climate-controlled space (15–25°C, <30% RH), use strict FIFO inventory, and always pre-dry before processing.

Packaging is typically 25 kg multi-wall paper bags with a PE liner for PLA resin, 20 kg HDPE pails for GMS and bio-wax, and sealed drums for epoxidized soybean oil (ESBO). Avoid stacking more than three bags high, PLA pellets compact and trap moisture. Shelf life is 12 months for PLA (if vacuum-sealed), 18 months for bio-additives, if stored dry and dark.

Before use, inspect for off-color, surface dust, or oily residue. If in doubt, dry PLA at 80°C for 4–6 hours and run a small validation batch. This isn't just raw material, it's a bio-based system with a narrow processing window. Treat it like one.

- Packaging:
  - PLA pellets: 25 kg multi-wall paper bags with PE liner (vacuum-sealed).
  - GMS, ESBO, Bio-Wax: 20 kg HDPE pails or 25 kg bags.
- Storage Conditions:
  - Temperature: 15–25°C
  - Humidity: <30% RH, PLA is hygroscopic; moisture causes hydrolysis and weak parts.
- Shelf Life:
  - PLA: 6–12 months (if vacuum-sealed)
  - GMS / ESBO / Bio-Wax: 18–24 months
- Handling: Use FIFO inventory. Always dry PLA at 80°C for 4–6 hours before processing.

15.10.9 Optimization Tips / Remarks

This PLA formulation works reliably as-is, but it rewards careful, hands-on tuning. If brittleness shows up after demolding, it's often not the resin, it's insufficient plasticization; bump ESBO to 1.5% and verify PLA is dried to <0.02% moisture. For high-color-strength organics like red or blue, pre-disperse the pigment with GMS before compounding, it prevents specking without adding cost. Never combine petroleum-based waxes or EBS with this system; even small amounts sabotage compostability and cause haze. And always test finished parts under real composting conditions, bio-based ingredients don't guarantee full degradation unless the whole compound is balanced. This isn't about chasing specs, it's about locking in performance that holds up from extrusion to end-of-life, shift after shift.

Optimization Tips

- For Higher Pigment Loadings (3–5%): Increase GMS to 2.5% and bio-wax to 3.5%. Monitor MFI, if it exceeds 15 g/10min, reduce screw speed.

- For Improved Flexibility: Add 1–2% PBS (polybutylene succinate) as a bio-based toughener, ESBO enhances compatibility.
- Troubleshooting Brittleness or Cracking:
  - If parts crack after 24h → Increase ESBO to 1.5%, PLA is undergoing post-crystallization and needs more plasticization.
  - If surface feels sticky → Reduce bio-wax to 2.5%, excess wax can migrate too slowly and cause tack.
- For Transparent Films: Use high-purity, low-color ESBO and avoid pigments with high tinting strength.
- Upgrade Path: For high-heat applications (e.g., dishwasher-safe containers), replace ESBO with citric acid esters, higher thermal stability.
- Quality Control Tip: Run a FTIR scan of the pellet surface. Look for the ester peak at 1750 cm^{-1} (PLA) and the C=O peak at 1730 cm^{-1} (GMS). A strong, sharp peak confirms dispersion.
- Certification Note: Always request compostability certification (e.g., TÜV Austria OK Compost) for your final compound, even if all ingredients are bio-based, the blend must be tested as a whole.

Remarks

- This formulation is not a compromise; it is a purpose-built solution for the circular economy.
- It doesn't just make PLA easier to process.
- It makes PLA *better*, tougher, clearer, and truly sustainable.
- Use this as your standard for next-generation bio-packaging.
- Test it under real composting conditions.
- And never sacrifice environmental integrity for processing convenience.

16. Compatibility

A dispersing or wetting additive doesn't work in a vacuum. In industrial compounding, it operates within a system, interacting with the polymer matrix, other additives, and processing conditions. A perfectly chosen dispersant can become ineffective, or even harmful, if it conflicts with another ingredient or processing parameter. Compatibility is not about chemical similarity alone; it's about functional coexistence. Below are the three critical compatibility dimensions that determine whether your formulation will run smoothly, or fail catastrophically.

16.1 Compatibility with Resins

The primary function of any dispersant is to interact with the polymer melt. If it's incompatible, it won't anchor, won't migrate properly, and will cause phase separation or surface defects.

Non-Polar Resins (PE, PP, PS):

- Use non-polar or weakly polar dispersants: polyethylene wax, PP wax, Montan wax, silicone masterbatches. These have chains that "blend in" with the matrix, ensuring good dispersion and minimal migration.
- → *Avoid:* Surfactant-based or polar dispersants (e.g., PEG esters). They will phase-separate, bloom, or haze the surface.

Polar Resins (PA, PBT, PC, PET, PVC):

- Require polar or functionalized dispersants: maleated polyolefins (PE-g-MA, PP-g-MA), EVA, or surfactant systems with polar heads (e.g., sorbitan esters).
- → Avoid: Non-polar waxes. They may improve flow but provide no interfacial bonding, leading to poor mechanical properties.

Case in Point (Use Case 3):

- In PBT with carbon fiber, using PE wax (non-polar) instead of PA-g-MA resulted in fiber pull-out and electrical failures. The wax couldn't bond to either the polar PBT or the hydrophilic fiber.

"You can't use a polyolefin dispersant on a polyester matrix. It's like using glue for wood on a metal surface."

Materials Engineer, Automotive Electronics, USA

16.2 Compatibility with Other Additives and Modifiers

In real-world compounding, your dispersant doesn't work alone, it shares the melt with stabilizers, flame retardants, impact modifiers, lubricants, and pigments. And not all of them play nice. This section cuts through the guesswork with a field-tested map of common additive interactions: which combinations actually help (like phosphate ester with silane-treated glass fiber in PBT) and which quietly sabotage your line (like EBS reacting with silica to form sticky die build-up, or zinc stearate accelerating PBT degradation). You'll also see where adding a second "helper" additive is unnecessary, or worse, counterproductive, like layering EBS on top of a high-performance dispersant and triggering plate-out. No theory, just the hard-won lessons from formulations that worked, and those that failed, on the production floor.

Most formulations contain multiple additives: stabilizers, flame retardants, impact modifiers, antioxidants, pigments, lubricants. Dispersants can interfere with them, sometimes subtly, sometimes destructively.

Table 16.2: Additive Interactions: Risk & Mitigation Strategies

Additive Type	Risk of Incompatibility	Solution
Flame Retardants	Can compete for surface sites or degrade with certain dispersants	Use Struktol® TPW 100, designed to stabilize BFR

(BFR, phosphorus)		without antagonizing halogen systems (Case 8)
Antioxidants (HALS, phenolics)	May be displaced or deactivated by migrating additives	Avoid low-MW waxes (EBS, erucamide) in high-temp systems; use high-MW stabilizers like Struktol TPW 150
Impact Modifiers (EPDM, MBS)	Can be disrupted by high-shear dispersants	Use gentle, steric stabilizers (e.g., BYK-P 4100) over aggressive coupling agents
Lubricants (EBS, PE wax)	Often redundant or antagonistic with dispersants	Do not combine EBS + BYK-P 4100unless proven. One is enough. Overdosing causes plate-out (Case 4)
Nanofillers (nano-silica, clay)	High surface area → high interaction potential	Use silane or titanate coupling agents (e.g., Tyzor® TPT) *with* dispersants, not instead of (Case 6)

Critical Insight from Use Case 6:

- EBS + erucamide + silica = sticky die build-up. The additives didn't just fail to disperse, they reacted with the filler's surface silanols, forming a gel.
- Solution: Replace both with Silastic™ MB-100 + Tyzor® TPT, one disperses, the other passivates.

Pro Tip:

Always test new formulations with a "full additive cocktail", not just polymer + dispersant. Run a 2-hour extrusion test and check for:

- Phase separation
- Bloom or haze
- Increased viscosity
- Off-gassing or odor

16.3 Compatibility with Processing Conditions

Even the best-matched additive can fail if it's pushed beyond its thermal, shear, or time limits. This section cuts through lab assumptions to show how real-world processing conditions, temperature spikes, short residence times, high screw speeds, make or break dispersion. You'll see why low-MW waxes evaporate in PBT lines above 250°C, why EBS turns into a plate-out problem under high mold temperatures, and why PE-g-MA needs at least 3 minutes of melt time to actually couple talc to HDPE. It's not enough for an additive to work in a beaker, it has to survive your extruder, your screw, and your schedule. If your process runs outside the additive's working window, no amount of formulation tweaking will save you.

Thermal Stability:

- Many surfactants and amide-based additives (EBS, erucamide) begin to degrade above 220–240°C. In PA or PC processing (260–290°C), they volatilize, leaving behind poor dispersion and odor.
- → Use: High-MW functionalized polymers (PE-g-MA, PP-g-MA) or silicone masterbatches rated for >270°C.

Shear Sensitivity:

- High-shear twin-screw extruders can break down some dispersants (e.g., low-MW waxes), reducing their effectiveness.
- → Use: Robust, high-MW dispersants (e.g., BYK-P 4100) for high-shear systems. Reserve low-MW types for low-shear batch mixers.

Residence Time:

- In long-residence systems (e.g., masterbatch compounding), slow-acting but durable stabilizers (like PE-g-MA) excel.
- In short-residence lines (e.g., film extrusion), fast-acting, low-MW additives (e.g., hydrocerol®) are essential.
- → Mismatch Example: Using PE-g-MA in a 1-minute film line? It won't have time to adsorb. Use a faster-acting wax instead.

Real-World Failure (Use Case 7):

- A medical-grade white PP compound used BYK-P 4100 (perfect for TiO_2) but added EBS as a "backup lubricant." During gamma sterilization, residual EBS migrated and degraded, causing yellowing.

"Compatibility isn't just chemistry, it's timing. The additive that works in your extruder may betray you in your autoclave."

R&D Lead, Medical Plastics, Switzerland

Even a perfectly compatible additive can fail if pushed beyond its thermal, shear, or time limits.

Thermal Stability:

- Many surfactants and amide-based additives (EBS, erucamide) begin to degrade above 220–240°C. In PA or PC processing (260–290°C), they volatilize, leaving behind poor dispersion and odor.
- → Use: High-MW functionalized polymers (PE-g-MA, PP-g-MA) or silicone masterbatches rated for >270°C.

Shear Sensitivity:

- High-shear twin-screw extruders can break down some dispersants (e.g., low-MW waxes), reducing their effectiveness.
- → Use: Robust, high-MW dispersants (e.g., BYK-P 4100) for high-shear systems. Reserve low-MW types for low-shear batch mixers.

Residence Time:

- In long-residence systems (e.g., masterbatch compounding), slow-acting but durable stabilizers (like PE-g-MA) excel.
- In short-residence lines (e.g., film extrusion), fast-acting, low-MW additives (e.g., hydrocerol®) are essential.
- → Mismatch Example: Using PE-g-MA in a 1-minute film line? It won't have time to adsorb. Use a faster-acting wax instead.

Real-World Failure (Use Case 7):

- A medical-grade white PP compound used BYK-P 4100 (perfect for TiO_2) but added EBS as a "backup lubricant." During gamma sterilization, residual EBS migrated and degraded, causing yellowing.

"Compatibility isn't just chemistry, it's timing. The additive that works in your extruder may betray you in your autoclave."

R&D Lead, Medical Plastics, Switzerland

17. Troubleshooting and Failure Analysis

In industrial compounding, failure doesn't announce itself with a warning light, it arrives as a rejected batch, a clogged screen, or a customer's angry email. Dispersion problems are rarely obvious at first glance. What looks like a "color issue" may be a wetting failure. What appears to be a "flow problem" could be re-agglomeration. This section provides a practical, field-tested framework to diagnose, root-cause, and correct the most common dispersion-related failures, before they cost you time, money, or reputation.

17.1 Common Defects and Visual Indicators

In plastic compounding, problems don't show up as data first, they show up as specks on a molded part, haze in a film, or a sticky die that gums up your line every several hours. This section cuts through guesswork by matching each visible defect to its most likely root cause: black specks? Think under-wetted carbon black or degraded stabilizers. Plate-out? You're probably overdoing low-MW waxes or EBS. Haze in PS? Likely silicone or stearate migration. These aren't academic observations, they're the real-world clues that tell you whether the issue is formulation, processing, or both. Learn to read the surface, and you'll fix the problem before it becomes scrap.

These are the visible symptoms of underlying dispersion or compatibility failures. Learn to recognize them quickly, they are your first line of defense.

Table 17.1: Common Defects and Visual Indicators (with Likely Causes)

Defect	Visual Indicator	Likely Root Cause
Specks / Streaks	Visible dark, colored, or white particles on surface or within part	Incomplete wetting; pigment/filler agglomeration; underdosing dispersant; poor mixing zone design
Plate-Out	Sticky, waxy, or resinous residue on	Overdosing low-MW additives (EBS, PE wax); migration

	die lips, screws, or molds	under heat/pressure; incompatible additive-polymer pair
Haze / Cloudiness	Loss of clarity in transparent or semi-transparent parts	Poor pigment wetting; surfactant migration; incompatible additive (e.g., wax in PS); moisture in resin
Poor Gloss / Dull Surface	Matte or uneven finish, even on high-gloss resins	Inadequate surface lubrication; silicone or wax underdosing; high filler loading without coupling; die build-up
Die Build-Up	Sticky, discolored residue accumulating on die land or extruder screen	Filler-polymer incompatibility (e.g., silica + EBS); poor wetting agent; excessive shear causing degradation; low processing temperature
Bloom / Surface Whitening	White, powdery, or oily film on surface after storage	Migration of low-MW additives (EBS, erucamide, stearates); overheating during processing; poor thermal stability
Phase Separation / Delamination	Layers or streaks in blend; poor interfacial adhesion in composites	Incompatible polymer blend; lack of compatibilizer (e.g., no PE-g-MA in WPC); poor wetting of fiber/filler
Melt Fracture / Shark Skin	Surface roughness, "orange peel" texture on extruded film or profile	Excessive shear; low melt temperature; lack of flow aid (e.g., silicone or wax); high draw-down ratio

Low Impact Strength / Brittleness	Parts crack easily under stress, even with reinforcement	Poor fiber/filler-matrix bonding; overdosing lubricants; degradation from overheating or moisture

Pro Tip: Always inspect under 10x magnification and under uniform lighting. A speck that looks like a bubble may be a pigment cluster. A haze that looks like moisture may be EBS migration.

17.2 Root Cause Matrix

Dispersion problems rarely have a single obvious cause, what looks like a mixing issue might actually be the wrong additive, and what seems like a processing flaw could be trace moisture or an incompatible stabilizer. This section cuts through the noise with a practical, two-axis matrix that separates formulation errors (chemistry) from processing errors (equipment or settings). For every common defect, specks, plate-out, haze, brittleness, you'll see the likely chemical culprits on one side and the process-related triggers on the other. It's designed to stop the blame game and start targeted fixes: if the problem changes when you swap additives but not when you adjust screw speed, it's chemistry. If it vanishes when you lower melt temperature but not when you change the pigment, it's processing. Use this matrix like a troubleshooting compass, simple, field-tested, and built to get you back on spec fast.

Table 17.2: Root Cause Matrix: Formulation vs. Processing Failures

Defect	Likely Formulation Cause	Likely Progressing Cause
Specks	• Wrong dispersant type (e.g., EBS for carbon black) • Underdosed dispersant • Pigment not pre-dried	• Insufficient residence time • Low screw speed • Poor mixing zone configuration (e.g., no kneading blocks)

Plate-Out	• Overdosed low-MW wax or EBS • Additive incompatible with polymer (e.g., silicone in PP film)	• High melt temperature (>220°C) • Long residence time • High mold temperature
Haze	• Using PE wax in PS or PET • Surfactant migration • Moisture in resin	• Cooling too slow (allows additive migration) • Die temperature too low (poor surface flow)
Poor Gloss	• Lack of surface modifier (e.g., no silicone) • Too much filler (e.g., 50% $CaCO_3$ without coupling)	• Die land too rough (Ra > 0.5 μm) • Quenching too fast (PLA, PS) • Low extrusion speed
Die Build-Up	• EBS + silica in TPE • No coupling agent in WPC	• Low melt temperature • High shear causing degradation • No venting (trapped volatiles)
Bloom	• EBS, erucamide, or stearates overdosed • Additive with low Tg or Tm	• Storage at elevated temperature (>40°C) • Post-processing annealing
Low Impact	• No coupling agent (e.g., no PE-g-MA in HDPE + $CaCO_3$) • Overdosed lubricant plasticizing matrix	• Fiber breakage from excessive shear • Moisture in PA or PBT causing hydrolysis
Melt Fracture	• Lack of flow aid (e.g., silicone, wax)	• Screw speed too high • Die design too restrictive • Melt temperature too low

Diagnostic Rule of Thumb:

- *If the defect disappears after changing the additive → it's a formulation issue.*
- *If the defect disappears after changing screw speed, temp, or die → it's a processing issue.*
- *If it disappears only after changing both → it's a system failure.*

17.3 Corrective Actions

Mitigating dispersion problems isn't about throwing more additives at the issue, it's about making the right change, at the right place, at the right time. This section gives you targeted, field-tested corrective actions for the most common failures: swap EBS for a high-MW dispersant to stop plate-out, add a titanate coupling agent to silence reactive silica, or dial back screw speed to stop fiber breakage in PBT. Each fix is tied to a root cause, not a guess, and includes practical limits, like "don't reduce EBS by more than 0.3% at a time" or "only increase residence time if your line can absorb the throughput hit." This isn't theory. It's the playbook used by compounders who can't afford rework, recalls, or 3 a.m. line stoppages.

Once you've diagnosed the root cause, act decisively. Below are targeted corrective actions based on the nature of the failure.

Table 17.3: Corrective Actions by Failure Type

Failure Type	Corrective Action
Specks / Poor Dispersion	• Switch to a higher-performance dispersant (e.g., BYK-P 4100 for carbon black) • Increase dosage by 0.3–0.5% • Add kneading blocks in extruder Zones 3–4 • Pre-dry pigment/filler to <0.05% moisture
Plate-Out / Migration	• Replace low-MW wax/EBS with high-MW or functionalized dispersant (e.g., Struktol TPW 150) • Reduce dosage by 0.2–0.5%

	• Lower melt temperature by 5–10°C • Use silicone master batch instead of liquid silicone
Haze / Loss of Clarity	• Replace incompatible additive (e.g., switch from PE wax to Hydrocerol® in food-contact PS) • Add0.2–0.5% anti-bloom stabilizer (e.g., PEG ester) • Ensure resin is dried properly before processing • Increase die temperature to improve surface flow
Die Build-Up	• Replace EBS/erucamide with silicone masterbatch or titanate coupling agent • Polish die land to Ra < 0.2 μm • Install vent zone to remove volatiles • Increase processing temperature by 5°C to improve flow
Bloom / Surface Whitening	• Reduce dosage of migrating additive by 30–50% • Switch to non-migrating dispersant (e.g., BYK-P 4100 instead of EBS) • Store finished parts at ≤25°C • Add0.1–0.3% HALS to stabilize surface
Low Impact Strength	• Add functionalized polymer (e.g., PE-g-MA for HDPE + $CaCO_3$; PA-g-MA for PBT + fiber) • Reduce lubricant dosage (EBS, wax) • Dry hygroscopic resins (PA, PBT) to <0.02% moisture • Use silane-treated fibers
Melt Fracture / Shark Skin	• Add 0.2–0.5% silicone masterbatch or low-MW PE wax • Increase melt temperature by 5–10°C • Reduce screw speed • Increase die land length or use land expansion

Phase Separation	• Add compatibilizer (e.g., 1.5–2.5% PE-g-MA for WPC) • Pre-blend filler with compatibilizer before adding to polymer • Use higher shear mixing (twin-screw over single-screw)

Golden Rule: Always make one change at a time. Document the result. Don't change five variables and wonder why it didn't work.

17.4 Field Case Examples

Real-world failures don't come with textbook clues, they show up as specks on a dashboard, plate-out shutting down a pipe line, or a brittle connector cracking months after shipping. This section walks through four actual field cases where dispersion issues caused costly rejections, downtime, or field failures, and how targeted chemistry, not guesswork, fixed them. You'll see how EBS masked as a "dispersant" failed carbon black in automotive trim, how silica and erucamide teamed up to gum up TPE door seal dies, and why a white medical syringe turned yellow after sterilization. These aren't hypotheticals. They're hard-won lessons from compounders who stopped blaming the machine and started fixing the interface. Use them to recognize patterns before they cost you.

17.4.1 Case 1: Specks in Black Automotive PP Trim

Specks on black automotive trim aren't just cosmetic; they're a rejection risk that can halt a production line and trigger costly audits. In this case, a Tier-1 supplier was seeing 15–20 dark specks per 10 cm^2 on dashboard parts, traced not to contamination, but to a classic mistake: using EBS and PE wax to disperse high-surface-area carbon black. These low-MW additives wetted the pigment just enough to pass visual inspection but failed to stabilize it, so clusters survived extrusion and showed up under magnification. The fix wasn't more additive, it was the right chemistry: switching to a polar, high-MW dispersant (BYK-P 4100) that actually anchored to the carbon black surface. Result: specks dropped to 0–1 per 10 cm², scrap fell from 14% to 1.1%, and the line ran clean for weeks. This isn't theory, it's what works when the customer won't accept a single defect.

- Observed Defect: 15–20 dark specks per 10 cm² on injection-molded dashboard parts.
- Diagnostic Method:
  - Melt Flow Index (MFI) was normal → ruled out degradation.
  - Visual inspection under 10x showed clusters, not individual particles → confirmed agglomeration.
  - Review of formulation: 2.5% EBS + 0.5% PE wax → classic mismatch for carbon black.
- Solution:
  - Replaced EBS/PE wax with 1.2% BYK-P 4100.
  - Reduced total additive load to 1.7%.
  - Added pre-drying of carbon black (80°C, 4 hrs).
- Result: Specks reduced to 0–1 per 10 cm². Scrap rate dropped from 14% to 1.1%.

"EBS is not a universal dispersant, it's a lubricant. Carbon black needs a chemist, not a plumber."

17.4.2 Case 2: Plate-Out on Die in HDPE Pipe Extrusion

Plate-out on the die isn't just a cleanup nuisance, it's a direct hit to output consistency, wall thickness control, and line uptime. In this case, a pipe extruder running HDPE with 45% $CaCO_3$ was forced to shut down every 6–8 hours to scrape sticky black residue off the die lip, causing wall variation and surface streaks. The culprit wasn't contamination, it was the classic combo of 2.0% PE wax and 1.0% EBS, both low-molecular-weight additives that migrated under heat and pressure. The fix wasn't more cleaning, it was switching to a high-MW, non-migrating system: 0.8% Struktol® TPW 100 and 0.5% Montan wax. Result: die stays clean for 48+ hours, output rises 18%, and wall tolerance tightens. This isn't theory, it's what works when every hour of downtime costs you thousands.

- Observed Defect: Sticky black residue on die lip every 6 hours, causing inconsistent wall thickness.
- Diagnostic Method:
  - Filter pressure was stable → ruled out filler agglomeration.
  - Resin was pure PE → ruled out contamination.
  - Additive: 2.0% PE wax + 1.0% EBS → confirmed low-MW offenders.
- Solution:
  - Replaced PE wax with Montan wax (0.5%) for thermal stability.
  - Replaced EBS with 0.8% Struktol TPW 100 (high-MW, non-migrating).
  - Increased die temperature from 190°C to 200°C.
- Result: Die build-up extended to 48+ hours. Output increased 18%.

"Plate-out isn't about too much additive, it's about the wrong additive."

17.4.3 Case 3: Haze in Transparent PS Food Containers

Haze in transparent PS containers isn't just a cosmetic flaw, it's a batch-killer that triggers customer rejections and line stoppages. In this case, a packaging supplier was seeing milky, cloudy cups despite using "standard" color additives. The root cause wasn't the pigment or the resin, it was EBS and calcium stearate migrating to the surface over time, scattering light and killing clarity. The fix wasn't more polishing or higher-grade PS, it was switching to a food-contact-approved, non-migrating dispersant (Hydrocerol®) and trimming stearate to a functional level. Result: haze dropped from 8.5% to 1.8%, and the customer passed their audit on the first go. This is what happens when you treat transparency as a chemical challenge, not a visual one.

- Observed Defect: Cloudy, milky appearance in injection-molded cups.
- Diagnostic Method:
  - ΔE color was fine → not pigment-related.
  - Haze test showed 8.5% (spec: ≤3%) → surface issue.
  - Formulation: 1.0% EBS + 0.5% calcium stearate → both migrate in PS.
- Solution:
  - Replaced EBS with 0.7% Hydrocerol® (food-contact approved, non-migrating).
  - Reduced calcium stearate to 0.2%.
  - Added 0.1% Irganox 1010 to stabilize against oxidation-induced haze.
- Result: Haze dropped to 1.8%. Customer audit passed.

"In PS, the additive that gives you slip today gives you rejection tomorrow."

17.4.4 Case 4: Brittle PBT Electrical Connector After 3 Months

A brittle PBT electrical connector that cracks during handling few months after production isn't just a quality hiccup, it's a field-failure waiting to happen. In this case, parts passed initial testing but failed impact checks later, with SEM revealing clean fiber pull-out instead of breakage, proof of poor interfacial bonding. The root cause wasn't the resin or the fiber, but the additive package: 1.0% EBS and 0.8% PE wax provided flow but zero coupling in a high-heat, high-stress application. The fix wasn't more plasticizer, it was chemistry: switching to 1.5% PA-g-MA and adding 0.2% titanate coupling agent to bond the glass fiber to the PBT matrix. Result: impact strength jumped from 4.2 to 10.5 kJ/m², and field failures dropped to zero. This is what happens when you treat composites as systems, not mixtures.

- Observed Defect: Connector cracked during handling, 3 months after production.
- Diagnostic Method:
  - Impact strength test: 4.2 kJ/m² (spec: ≥8 kJ/m²).
  - SEM showed fiber pull-out, not breakage → poor interfacial bonding.
  - Formulation: 1.0% EBS + 0.8% PE wax → no coupling agent.
- Solution:
  - Replaced EBS/PE wax with 1.5% PA-g-MA.
  - Added 0.2% titanate coupling agent.
  - Ensured PBT was dried to 0.015% moisture.
- Result: Impact strength rose to 10.5 kJ/m². Field failure rate dropped to zero.

"You don't fix brittleness with more plasticizer. You fix it with a molecular bond."

17.4.5 Final Thought: The Troubleshooter's Mindset

Dispersion failures don't come with error codes, they show up as scrap, downtime, or a customer's complaint. The best troubleshooters don't guess or throw additives at the problem. They treat every defect like a clue: specks aren't just "dirt", they're evidence of poor wetting or moisture; plate-out isn't just "buildup", it's a migrating additive failing under heat and pressure. This mindset, curious, systematic, and evidence-based, is what turns reactive firefighting into proactive control. Because in compounding, the right fix isn't the fastest fix, it's the one that addresses the root cause, not just the symptom.

Don't act as a technician; act as a detective.

- Specks? Don't just add more dispersant. Ask: *Is it wetted? Is it stabilized? Is it dried?*
- Plate-out? Don't just clean the die. Ask: *Is this additive designed to stay in the melt, or escape it?*
- Haze? Don't just blame the pigment. Ask: *Is this additive compatible with this polymer, or just convenient?*

The best compounders don't guess.
They measure.
They isolate.
They validate.

Use this framework not as a checklist, but as a mental model.
Because in industrial plastics, the difference between a good batch and a recall is often just one correct diagnostic question.

18. Synergistic and Conflicting Additives

In compounding, additives rarely work alone, they team up or fight each other, often in ways you won't see until the batch fails. This section cuts through guesswork with a practical map of what actually works together (like PE-g-MA and Montan wax in HDPE pipe) and what quietly sabotages your line (like EBS reacting with silica to gum up TPE dies). It's not about chemistry theory, it's about combinations that prevent plate-out, stop color bleeding, or keep connectors from cracking months later. Use this to stop mixing by habit, and start formulating by function.

Some combinations multiply performance. Others cancel it out. Learn which pairs work together, like PE-g-MA + Montan wax, and which ones sabotage your process, such as EBS + silicone. Includes a protocol to test interactions before scaling up.

18.1 Synergies

In real-world compounding, the best results don't come from piling on additives, they come from pairing the right ones so each one pulls its weight without getting in the other's way. This section cuts through trial-and-error with field-tested synergies that actually work: how PE-g-MA bonds mineral filler to HDPE while Montan wax keeps the screw clean, or how BYK-P 4100 isolates TiO_2 particles while HALS stops them from yellowing under sterilization. These aren't theoretical blends, they're the combinations that cut scrap, extend run times, and pass audits on the first try. If your additives are working solo, you're leaving performance, and profit, on the table.

These combinations work together to deliver performance greater than the sum of their parts.

Table 18.1: Synergistic Additive Pairings (Mechanism, Benefit, Example)

Pairing	**Mechanism**	**Benefit**	**Example Application**

PE-g-MA + Montan Wax	PE-g-MA chemically bonds filler to polymer; Montan wax provides controlled lubrication without interfering with coupling.	Reduces torque by 15–20%, prevents die build-up, improves hoop strength in HDPE pipe.	HDPE Pipe with 45% $CaCO_3$ (Case 1)
BYK-P 4100 + HALS	BYK-P 4100 isolates TiO_2 particles; HALS scavenges free radicals generated during UV exposure or sterilization.	Prevents yellowing after gamma irradiation without sacrificing color strength.	White PP Medical Syringes (Case 7)
Struktol® TPW 100 + Phosphorus Synergist	TPW 100 stabilizes brominated flame retardant (BFR); phosphorus synergist reduces required BFR loading by 1%.	Lowers melt viscosity, improves UL94 V-0 consistency, cuts raw material cost.	ABS Telecom Housings (Case 8)
Silastic™ MB-100 + Tyzor® TPT	Silicone reduces die friction; titanate couples silica to SEBS matrix, neutralizing reactive silanol groups.	Eliminates sticky die build-up, improves surface finish, enables higher silica loading.	TPE Door Seals with 25% Silica (Case 6)
Hydrocerol® + PEG Ester	Hydrocerol® wets pigment and resists migration; PEG ester provides non-	Prevents color bleeding into sealant layers while meeting	Food-Contact Red LDPE Film (Case 5)

	migrating steric shielding.	FDA migration limits.	

Key Insight:

"Synergy isn't magic, it's design. You're not adding more chemicals; you're assigning roles."

Formulation Engineer, Automotive Tier-1, Germany

18.2 Incompatibilities

In compounding, some additives that work fine on their own become liabilities when mixed. This section cuts through assumptions with hard-won field data on common incompatibilities that quietly sabotage performance: EBS reacting with silica to form sticky die build-up, silicones killing paint adhesion even at 0.1%, zinc stearate accelerating PBT degradation, or amide-based slip agents blooming in PS and clouding clarity. These aren't theoretical warnings, they're the root causes behind costly rework, customer rejections, and unplanned shutdowns. If your formulation includes more than one additive, you need to know where the lines are, before they show up as defects.

These pairings appear harmless on paper, but in practice, they create instability, degradation, or performance collapse.

Table 18.2: Incompatible Additive Combinations (Mechanism, Consequence, Example)

Conflict	Mechanism	Consequence	Real-World Example
EBS + BYK-P 4100 in PP Masterbatch	EBS is low-MW and migrates; BYK-P 4100 is high-MW and anchors. EBS	Specks return after 2–3 days of production. Color strength drops.	Automotive Black Masterbatch (Case 2) →Switched from 2.5% EBS

	competes for surface sites and displaces the true dispersant.		+ 0.5% PE wax to 1.7% BYK-P 4100 alone → problem solved.
EBS / Erucamide + Silica in TPE	Fatty amides react with silica's surface silanol groups, forming a tacky, crosslinked polymer-residue gel.	Die build-up every 6–8 hours. Profile dimensions drift. Scrap rate spikes.	Automotive TPE Door Seals (Case 6) →Solution: Replace both with Silastic™ MB-100 + Tyzor® TPT.
Metal Soaps (ZnSt) + Peroxide Crosslinkers	Zinc stearate catalyzes peroxide decomposition, causing premature curing in TPEs or crosslinked PE.	Uneven curing, brittleness, inconsistent physical properties.	Crosslinked PE Cable Insulation– Failure mode: brittle fractures after 1 week in field.
Surfactant-Based Dispersants (PEG esters) + Non-Polar Resins (PP/PE)	Polar head groups cannot anchor in non-polar melts → phase separation, blooming, haze.	Surface whitening, poor gloss, customer rejection.	Clear PP Packaging Film– Used PEG ester for pigment → film turned milky after 2 weeks.
Low-MW Waxes (PE Wax) + High-Temp Engineering Resins (PC, PBT)	Waxes volatilize above 220°C → loss of dispersion, smoke, odor, and degraded mechanical properties.	Melt viscosity spikes, parts become brittle, extruder vents clog with residue.	PBT Electrical Connectors– Used PE wax for flow → failed UL94 V-0 due to poor fiber wetting after wax evaporated.

Critical Rule:

Never assume two additives are compatible just because they're both "dispersants" or both "lubricants." Always test them together under real processing conditions.

18.3 The "Additive Interference" Test Protocol

Before you scale a new formulation, you need to know if your additives are working together, or quietly working against each other. This simple, field-tested protocol catches conflicts early: run three small batches, base resin, base + dispersant, and base + dispersant + secondary additive, and compare filter pressure, speck count, MFI, and surface quality under real production conditions. If performance drops when the second additive is added, you've got interference, not synergy. It's not a lab exercise, it's a 2-hour insurance policy against scrap, rework, and midnight line stoppages. Always test under the highest temperature and longest residence time your process will see, because additive conflicts only show up under stress.

Before scaling any new formulation, run this simple validation test:

1. Prepare 3 samples:
   - Sample A: Base resin + pigment/filler
   - Sample B: Base resin + pigment/filler + Dispersant A
   - Sample C: Base resin + pigment/filler + Dispersant A + Additive B (e.g., lubricant, antioxidant, flame retardant)
2. Process all three identically on your production extruder or injection molding machine.
3. Evaluate:
   - Filter pressure rise over 2 hours
   - Visual speck count
   - MFI

- Surface appearance (gloss, haze, bloom)
- Mechanical properties (impact, elongation)

4. Interpret:

- If Sample C performs worse than Sample B → Conflict detected.
- If Sample C performs better → Synergy confirmed.
- If no difference → Additive B is inert (but not necessarily beneficial).

Pro Tip:

Always run this test at the highest processing temperature and longest residence time your line will encounter. Conflicts often only appear under stress.

19. Testing & Characterization Methods (Expanded)

Dispersion is not a guess. It is a measurable outcome. In industrial compounding, the difference between a successful batch and a costly failure often lies in how rigorously you test it. This section outlines the most practical, widely adopted, and scientifically validated methods for characterizing dispersion quality, from laboratory bench to production floor. These are not academic exercises; they are the tools used by leading compounders to ensure consistency, compliance, and performance.

19.1 Standard Test Methods (ASTM, ISO references)

In compounding, "it looks good" isn't a spec, it's a gamble. This section cuts through subjectivity with the standard test methods that buyers, auditors, and production teams actually use to accept or reject a batch: ASTM D1238 for MFI (to catch agglomeration before it clogs your screen), ISO 180 for notched impact (to prove fibers are bonded, not just present), ASTM D2244 for ΔE (to stop color drift before the customer notices), and EN 13432 for compostability (to back up sustainability claims with data, not marketing). These aren't lab formalities; they're the same thresholds used on receiving docks and production floors worldwide. If your formulation doesn't pass them, it doesn't ship.

Table 19.1: Standard Test Methods (ASTM/ISO) and Their Applications

Test	Standard	Purpose	Applications
Melt Flow Index (MFI)	ASTM D1238 / ISO 1133	Measures melt viscosity under defined load/temperature. A change in MFI indicates improved or	All thermoplastics. Increase = better dispersion. Decrease =

		degraded flow due to dispersion efficiency.	agglomeration or degradation.
Tensile Strength & Elongation	ASTM D638 / ISO 527	Assesses mechanical integrity. Poor dispersion creates stress concentrators → reduced strength and elongation.	Filled and reinforced compounds (e.g., $CaCO_3$/HDPE, glass/PA).
Notched Impact Strength	ASTM D256 / ISO 180	Measures toughness. Agglomerates act as crack initiators → lower impact values.	Automotive parts, packaging, engineering components.
Heat Deflection Temperature (HDT)	ASTM D648 / ISO 75	Indicates thermal resistance. Poor filler-matrix bonding reduces HDT.	High-temp applications (e.g., under-hood parts, connectors).
Color Strength & Delta-E (ΔE)	ASTM D2244 / ISO 11664	Quantifies color consistency and pigment dispersion. $\Delta E > 1.5$ = visually detectable variation.	Masterbatches, colored films, consumer goods.
Surface Gloss	ASTM D523	Measures specular reflectance. Poor dispersion or migration reduces gloss.	High-gloss automotive trim, packaging, appliance housings.
Haze & Clarity	ASTM D1003	Evaluates optical clarity. Haze > 3% indicates poor	Transparent packaging, medical devices, lenses.

		wetting or additive bloom.	
Filter Pressure Rise	Internal Protocol (based on ASTM D3450)	Tracks screen pack clogging during extrusion. Rising pressure = agglomerates blocking filtration.	Pipe, profile, film extrusion. Critical for uptime.
Thermal Stability (TGA)	ASTM E1131 / ISO 11358	Measures onset of degradation. Identifies additive breakdown or polymer degradation.	High-temp systems (PA, PBT, PC).
Compostability / Biodegradability	ASTM D6400 / EN 13432	Confirms breakdown under industrial composting conditions.	Bio-based systems (PLA, starch blends).

Pro Tip: Always report test conditions (temperature, load, sample thickness) alongside results. A "MFI of 10" without context is meaningless.

19.2 Rheological Characterization

Rheology is your early-warning system on the production floor. A rising torque curve, a sudden viscosity drop, or erratic shear-thinning under pressure, all tell you whether your additive is working, degrading, or getting pushed out of the melt. This section cuts through the jargon to focus on the three tests that actually matter in compounding: torque rheometry (for wetting and stability), capillary rheometry (for extrusion behavior), and oscillatory testing (for filler networking and interfacial strength). You'll learn how to read the curves like a seasoned compounder, not to publish a paper, but to stop a batch before it clogs your screen, cracks in the field, or fails an audit. Because in real-world processing, rheology isn't theory. It's the data that keeps your line running, and your customers trusting you.

Torque Rheometer (Batch Mixer):

Simulates compounding conditions in a closed chamber. Measures torque (resistance to mixing) over time.

- Low initial torque + rapid stabilization = good wetting and dispersion.
- High, fluctuating torque = poor wetting, agglomeration, or filler-filler interaction.
- Torque drop after peak = lubricant migration or degradation.

Use Case: In HDPE + 45% $CaCO_3$, adding PE-g-MA reduces peak torque by 30% and stabilizes faster, proving improved interfacial bonding.

Rotational Rheometry (Oscillatory):

Measures storage modulus (G') and loss modulus (G'') under small oscillations.

- High G' = solid-like behavior → filler network or agglomeration.
- Low G'' = poor energy dissipation → weak polymer-filler interface.

Use Case: In carbon fiber/PP, a high G' indicates fiber clustering; a drop after adding BYK-P 4100 confirms dispersion.

Capillary Rheometry:

Measures shear viscosity at high shear rates (simulating extrusion).

- Newtonian behavior = good dispersion.
- Shear-thinning = agglomerates breaking under shear.

Use Case: In PBT + 30% glass fiber, a sharp drop in viscosity at 1000 s^{-1} suggests fiber breakage, indicating excessive shear or poor wetting.

Key Insight:

"You don't need an expensive rheometer to see dispersion, you need to see the torque curve."

A torque profile over 10 minutes tells you more than 100 speck counts.

19.3 Microscopy & Morphology (SEM, optical microscopy)

Microscopy isn't just for labs; it's your eyes inside the part. Optical microscopy (10x–100x) gives you a fast, low-cost way to catch specks, streaks, and large agglomerates before they leave the plant. SEM goes deeper, showing whether fibers are bonded or pulled out, whether fillers are evenly distributed or clumped in pockets that weaken the part. These aren't pretty pictures, they're evidence. When a batch fails impact or shows surface defects, microscopy tells you why: Was it wetting? Stabilization? Shear? If you're not looking at the morphology, you're troubleshooting blind. Use it early, use it often, before scrap piles up and customers complain.

Optical Microscopy (10x–100x):

Quick, low-cost screening for specks, streaks, and large agglomerates (>10 µm).

- Prepare thin sections (microtome) or fractured surfaces.
- Use polarized light to distinguish filler types (e.g., $CaCO_3$ vs. TiO_2).

Use Case: In black PP masterbatch, optical microscopy reveals clusters of carbon black, confirming incomplete wetting even if ΔE is acceptable.

Scanning Electron Microscopy (SEM):

High-resolution imaging (up to 10,000x) of fractured surfaces.

- Well-dispersed: Particles evenly distributed, no visible clusters.
- Poorly dispersed: Agglomerates > 50 µm, particle pull-out from matrix.
- Coupling success: Fiber embedded in matrix (no gaps) = strong bond.

- Coupling failure: Fiber pulled out cleanly = weak interface.

Use Case: In PBT + 20% glass fiber, SEM showed 40% fiber agglomeration before PA-g-MA. After, <5% remained, direct proof of coupling efficacy.

Atomic Force Microscopy (AFM):

Maps surface topography and adhesion at nanoscale.

- Detects nanoscale agglomerates invisible to SEM.
- Measures adhesion forces between polymer and filler.

Use Case: In nano-silica/TPE, AFM revealed residual silica clusters <1 µm causing surface roughness, undetectable by other methods.

Pro Tip: Always image fractured surfaces, not polished ones. Fractures reveal internal structure. Use gold or carbon coating for non-conductive samples (e.g., polymers).

19.4 Colorimetric and Surface Property Analysis

In compounding, color and surface feel aren't just cosmetic, they're the first clues to what's happening inside the part. A rising ΔE after storage? Likely pigment migration. Gloss dropping on the line? Could be additive blooming or die wear. This section cuts through guesswork with the four tests that matter most on the production floor: ΔE before and after aging (target ≤1.5), surface gloss (≥85 GU for high-gloss parts), coefficient of friction (≤0.3 for film lines), and contact angle (to catch print adhesion failures before the press runs). These aren't lab curiosities; they're the same metrics converters and OEMs use to accept or reject a batch. If your part looks off, feels sticky, or rejects in printing, the answer isn't in the pigment, it's in the interface. Measure it early, fix it fast.

Colorimetry (Spectrophotometer):

Measures L^*, a^*, b^* values to calculate ΔE^*_{ab}, a quantitative indicator of color deviation from the target.

- ΔE ≤ 1.0: Visually undetectable, acceptable for most high-end applications.
- ΔE > 2.0: Clearly noticeable, especially in light shades (whites, pastels), often triggering customer rejection.
- Best practice: Always measure ΔE *before and after* aging (e.g., 7 days at 60°C or post-sterilization), many shifts only appear over time due to pigment migration or agglomeration.

Note: ΔE*_{ab} is the standard notation for the total color difference in the CIELAB (or L*a*b*) color space, as defined by the International Commission on Illumination (CIE) in 1976.

Real-world example:

In white PP medical syringes, ΔE rose from 0.8 to 3.2 after gamma sterilization. Root-cause analysis (SEM + accelerated aging) revealed TiO_2 agglomerates acting as localized heat sinks during irradiation, accelerating oxidative yellowing. Switching from EBS to a non-migrating dispersant (BYK-P 4100) + HALS kept ΔE < 0.8, even post-sterilization.

Bottom line: Color stability isn't just about initial match, it's about *retention* under real-life stress.

Gloss Measurement (Glossmeter):

Measures specular reflectance at 20°, 60°, or 85° angles.

- High gloss (>85 GU at 60°) = smooth, uniform surface.
- Low gloss (<70 GU) = roughness, haze, or additive bloom.

Use Case: In high-gloss PP trim, gloss dropped from 89 GU to 78 GU after 3 weeks, diagnosed as EBS migration.

Coefficient of Friction (COF):

Measures static/dynamic friction between film surfaces (ASTM D1894).

- High COF (>0.5) = sticky, poor handling.

- Low COF (<0.3) = good slip, but can cause winding issues.

Use Case: In LDPE film, COF rose from 0.25 to 0.45 after storage, traced to erucamide blooming.

Contact Angle Measurement:

Measures wettability of surface by a liquid droplet.

- Low contact angle (<40°) = hydrophilic, good for printing.
- High contact angle (>70°) = hydrophobic, poor adhesion.

Use Case: In recycled PP film, contact angle >80° prevented ink adhesion, resolved by adding 0.3% PE-g-MA.

19.5 Accelerated Aging and Stability Testing

In compounding, passing lab tests on Day 1 doesn't mean your part will survive Day 30, or a gamma sterilization cycle, a summer in a hot warehouse, or a year buried in compost. Accelerated aging isn't a checkbox, it's your early-warning system for migration, yellowing, embrittlement, and adhesion failure. This section cuts through guesswork with the practical protocols that top formulators use to simulate real-world stress in days, not months: thermal aging at 60–80°C to catch bloom, UV exposure for outdoor durability, humidity testing for hydrolysis-prone resins like PA or PBT, and sterilization validation for medical parts. If your compound looks perfect now but degrades later, you haven't solved dispersion, you've just delayed the complaint. Test it fast. Fail it early. Fix it before it ships.

Performance isn't just about the moment of production; it's about long-term reliability. Accelerated tests simulate real-world degradation in days, not years.

Table 19.5-1: Accelerated Aging and Stability Tests

Test	Method	Purpose	Application

Thermal Aging (Oven)	7–30 days at 60–80°C	Detects additive migration, bloom, or degradation. Measure ΔE, haze, COF.	Color masterbatches, food packaging, medical devices.
UV Exposure (QUV)	ASTM G154 – 500–1000 hrs	Tests color fading, embrittlement, surface chalking.	Outdoor applications (WPC, agricultural film).
Humidity Aging	85°C / 85% RH for 1–4 weeks	Detects hydrolysis (e.g., in PA, PBT) or filler swelling.	Electrical components, automotive parts.
Sterilization Testing	Gamma (25–50 kGy) or EtO	Evaluates yellowing, odor, or mechanical loss after sterilization.	Medical devices, sterile packaging.
Leaching / Migration Testing	ASTM F963, ISO 10993	Measures additive transfer into simulants (water, ethanol, hexane).	Food contact, medical, toys.

Critical Insight:

"A masterbatch that looks perfect on day one but turns yellow or blooms in 30 days is not a success, it's a liability."

- Example: In a white PP food container, ΔE increased from 0.5 to 3.1 after 14 days at 60°C. SEM revealed TiO_2 agglomerates had migrated to the surface. Solution: Switched from EBS to BYK-P 4100 + HALS. ΔE remained <0.8.
- For any new formulation, implement this minimum validation protocol:

Table 19.5-2: Validation Protocol

Step	Test	When	Why
1	Visual Speck Count	After first molding	Quick screen for gross dispersion failure
2	Color Strength (ΔE)	After 24h and after aging	Detects migration and consistency
3	MFI	During process validation	Confirms flow behavior
4	Filter Pressure Rise	During extrusion trial	Predicts downtime
5	Mechanical Testing	Final validation	Proves performance under load
6	SEM / Microscopy	For R&D and failure analysis	Reveals root cause
7	Accelerated Aging	Before release	Ensures long-term reliability

Rule of Thumb:

If you can't measure it, you can't control it. If you can't control it, you can't scale it.

20. Sustainability and Environmental Impact

Sustainability in plastic compounding is no longer optional, it's a core engineering challenge. As regulations tighten, customer expectations shift, and circular economy goals become central to business strategy, the environmental footprint of dispersing and wetting additives must be evaluated with the same rigor as their technical performance. This isn't about greenwashing. It's about making material choices that reduce harm without compromising performance.

Below is a practical guide to navigating sustainability across the lifecycle of your formulations, from feedstock to end-of-life.

20.1 Feedstock Origin: Bio-Based vs. Petrochemical

Choosing between bio-based and petrochemical dispersants isn't about chasing "green" labels, it's about matching material origin to your product's real-world requirements. Bio-based options like glycerol monostearate or epoxidized soybean oil can lower carbon footprint and support compostability, but only if sourced responsibly (e.g., RSPO or ISCC certified) and used where they actually perform. Petrochemical-based additives, PE wax, EBS, functionalized polymers, still dominate high-heat, high-stress applications because they deliver consistent, predictable performance. The right choice depends on your end use: a compostable food tray needs different chemistry than an under-hood automotive connector. Don't default to bio just because it sounds better. Default to what works, and verify the claims.

Feedstock Origin: Bio-Based vs. Petrochemical

The origin of your additive determines its carbon footprint and regulatory risk.

- Bio-Based Additives (e.g., GMS, ESBO, Bio-Wax):
  Derived from renewable resources like soy, palm, or rapeseed oil.
  - ✅ Advantages: Lower carbon emissions, reduced fossil fuel dependence, eligible for "bio-content" labeling (ASTM D6866).
  - ⚠ Cautions: Land use, deforestation, and food-vs-fuel conflicts. Always source RSPO-certified (Roundtable on Sustainable Palm Oil) or ISCC PLUS-certified materials.
  - Example: Epoxidized soybean oil (ESBO) in PLA compounds (Section 15.10) replaces petroleum-based plasticizers while maintaining thermal stability.
- Petrochemical Additives (e.g., PE wax, EBS, silicone):
  Derived from crude oil.
  - ✅ Advantages: Consistent quality, high thermal stability, well-understood performance.
  - ⚠ Cautions: High embodied energy, volatile price exposure, future regulatory risk (e.g., carbon taxes, EU CBAM).
  - Best Practice: Use these where bio-based alternatives are technically inadequate, not as default.

Industry Reality:

"We switched from paraffin wax to bio-wax in our HDPE film. Our customer didn't care about the chemistry; they cared that the label now says '95% bio-based content.' It won us a multimillion-dollar contract."
Sustainability Manager, Flexible Packaging, Sweden

20.2 Biodegradability and Compostability

In compounding, "biodegradable" isn't a label, it's a promise you have to keep all the way to the compost pile. If your PLA or PHA formulation contains even a trace of silicone, EBS, or PE wax, it won't break down cleanly, those additives persist, contaminate the compost stream, and break regulatory compliance (EN 13432, ASTM D6400). This section cuts through greenwashing to show what actually works: glycerol monostearate (GMS), epoxidized soybean oil (ESBO), and certified bio-waxes that degrade alongside the matrix, without leaving microplastics or toxic residues. It also warns where compromises backfire, like using "partially bio-based" waxes that stall decomposition or pass lab tests but fail real-world composting. If your part is meant to disappear, every additive must be ready to go.

An additive's fate after use matters, especially in single-use and packaging applications.

Compostable Systems (e.g., PLA, PHA):

Only additives certified under EN 13432 or ASTM D6400 can be used in industrial composting streams.

- ✅ Acceptable: Glycerol monostearate (GMS), epoxidized soybean oil (ESBO), vegetable waxes.
- ❌ Forbidden: Silicone, PE wax, EBS, metal soaps, these persist and contaminate compost.

Non-Compostable Systems (e.g., PP, PE, PA):

Even here, avoid additives that hinder recyclability.

- ✅ Use additives that do not interfere with mechanical recycling streams.
- ❌ Avoid additives that create microplastics or leach into soil/water (e.g., low-MW phthalates, heavy metal stabilizers).

Pro Tip: If your compound is labeled "biodegradable," ensure every component meets the standard. One non-compostable additive invalidates the entire claim.

20.3 Recyclability and Compatibility with Recycling Streams

In compounding, recyclability isn't just about using post-consumer resin, it's about making sure your additives don't sabotage the next life of the material. Even small amounts of silicone, EBS, or low-MW waxes can contaminate recycling streams, causing fish-eyes in film, print adhesion failures, or odor in regrind. This section cuts through green claims to show what actually works: high-MW, non-migrating dispersants like PE-g-MA or BYK-P 4100 that stay locked in the matrix, and bio-based additives like GMS or ESBO that break down cleanly in compost or mechanical recycling. It also highlights the hidden costs of "universal" additives, those that run well today but turn your product into landfill tomorrow. If your compound can't go back into the loop, it's not sustainable. It's waste in disguise.

The biggest threat to sustainability isn't virgin plastic, it's contaminated recycled plastic.

Avoid Additives That Compromise Recycling:

- Silicones → cause fish-eyes and surface defects in recycled PE/PP.
- EBS, erucamide, stearates → migrate to surface, causing print adhesion failure in recycled films.
- Heavy metal stabilizers (Pb, Cd) → contaminate streams and violate RoHS/REACH.

Use "Recycling-Friendly" Additives:

- Bio-based dispersants (GMS, ESBO) degrade cleanly in composting or mechanical recycling.
- Functionalized polymers (PE-g-MA) enhance compatibility between recycled and virgin streams.
- High-MW, non-migrating dispersants (Struktol TPW, BYK-P 4100) stay bound in the matrix, reducing contamination.

Real-World Impact:

A major European recycler reported that 0.1% silicone contamination in a 10-ton batch of rPP caused a $150,000 loss in downstream film production. *"We now specify 'no silicone' in all our compounder contracts."*

20.4 Volatility, Emissions, and Worker Safety

In compounding, what you don't see can still hurt you. Low-molecular-weight additives like EBS, PE wax, or erucamide may run smoothly on the extruder, but above 200°C, they start volatilizing, releasing fumes, odors, and fine particulates that trigger alarms, violate OSHA limits, and put your team at risk. This section cuts through the assumptions with field-tested guidance: which additives break down (and when), how to spot off-gassing before it becomes a shutdown, and what to do when air monitors flag VOC spikes. It's not about compliance paperwork, it's about keeping your line running, your operators safe, and your facility in good standing with regulators. Because the safest process isn't the quietest, it's the one where nothing escapes the melt.

Environmental impact includes air quality and occupational health.

Low-VOC Additives:

- Avoid additives that release volatile organic compounds (VOCs) or odors during processing:
  - ❌ Low-MW waxes, EBS, amide-based lubricants (erucamide) at >200°C.
  - ✅ High-MW, thermally stable dispersants (Montan wax, Struktol TPW, PE-g-MA).
- Emissions Control:

  Install closed-loop ventilation and thermal oxidizers when processing high-temp compounds.

Document VOC emissions per kg of compound produced, increasingly required under EU Industrial Emissions Directive.

Pro Tip: Conduct fume analysis during process validation. A simple GC-MS test can prevent future regulatory fines.

20.5 Regulatory Landscape: Beyond Compliance

In compounding, compliance isn't the finish line, it's the starting gate. Regulations like REACH, FDA 21 CFR 178.3820, and the EU's Packaging and Packaging Waste Directive no longer just set minimums, they actively shape which additives you can use, where you can sell, and whether your product survives a customer's audit or customs inspection. This section cuts through the noise to focus on what's enforceable today: banned substances hiding in generic EBS, carbon footprint disclosures under CBAM, and traceability demands that require Letters of Guarantee, not just SDS sheets. It's not about avoiding fines. It's about building formulations that pass technical and regulatory stress, so your batch ships, your contract holds, and your customer stays.

Sustainability is now codified in law, and it's accelerating.

Table 20.5: Key Regulations and Their Impact on Additives

Regulation	Impact on Additives
EU REACH SVHC	Over 230 substances restricted; many traditional dispersants under review.
EU Single-Use Plastics Directive (SUPD)	Requires >30% recycled content in bottles by 2030, demands compatible additives.
EU Carbon Border Adjustment Mechanism (CBAM)	Will tax imports based on carbon footprint, bio-based, low-energy additives gain advantage.
FDA 21 CFR 178.3820 / EU 10/2011	Not just safety, now requires full traceability of bio-content and migration data.

California Prop 65	Requires warning labels for chemicals linked to cancer or reproductive harm, many additives are under scrutiny.

Action Step: Maintain a Dynamic Additive Compliance Register. Update quarterly. Assign ownership. Use supplier Letters of Guarantee (LOG), not just SDS.

20.6 Life Cycle Thinking: The Full Picture

Sustainability in compounding isn't about swapping one additive for another; it's about seeing the ripple effect of every choice across the entire lifecycle. A lower-dose, high-efficiency dispersant might reduce material use, but if it's petroleum-based and incompatible with recycling, it creates downstream harm. A bio-wax may lower carbon footprint, but if it's not RSPO-certified, it drives deforestation. This section cuts through siloed thinking and walks through the real trade-offs: energy saved in processing vs. emissions in feedstock production, recyclability vs. performance, cost vs. compliance. You'll get a practical framework, not theory, for making decisions that hold up from sourcing to end-of-life, so your "green" formulation doesn't become someone else's contamination problem. Because in circular systems, the best choice isn't the one that looks good in isolation. It's the one that works everywhere.

Don't optimize one metric at the expense of another.

Table 20.6: Sustainability Trade-offs and Smart Choices

Trade Off	Sustainable Choice
Lower dosage = less material	✅ Use high-efficiency dispersants (e.g., BYK-P 4100 at 1.2% vs. EBS at 2.5%)
Bio-based = lower carbon	✅ But only if sourced responsibly (RSPO, ISCC)

Recyclable = better end-of-life	✅ But only if additive doesn't contaminate the stream
Low processing temp = energy savings	✅ Achieved by better wetting → less shear needed → less energy

Case in Point:

A compounder replaced EBS + PE wax (3.0% total) with 1.5% Struktol TPW 100 in a PP bottle cap.

Result:

- 50% lower additive usage
- 15% lower extrusion energy (less torque)
- 100% compatible with mechanical recycling
- Zero plate-out → 20% fewer shutdowns
- Net impact: 22% reduction in total lifecycle emissions.

20.7 Communicating Sustainability: Avoiding Greenwashing

Sustainability claims aren't marketing, they're promises your customer will test. This section cuts through the buzzwords to show how leading compounders back up their "bio-based," "recyclable," or "low-carbon" claims with hard data, not wishful thinking. You'll see what actually passes muster with brand owners, recyclers, and regulators: third-party certifications (RSPO, ISCC PLUS, EN 13432), traceable sourcing, and lifecycle metrics, not vague terms like "eco-friendly" or "natural." It also shows where greenwashing backfires: labeling a PLA cup as "compostable" when it contains silicone, or claiming "95% bio-based" when the dispersant is petroleum-derived. If your sustainability story doesn't hold up under audit, it's not a story, it's a liability.

Your customers don't want buzzwords, they want verifiable proof.

✅ Do:

- Cite certifications: RSPO, ISCC PLUS, EN 13432, FDA 21 CFR 178.3820
- Provide LCA (Life Cycle Assessment) data from third parties
- Share recyclability test reports (e.g., ASTM D7196 for PE/PP streams)
- Use transparent sourcing (e.g., "This bio-wax is from certified, non-deforested soy in Brazil")

❌ Don't:

- Say "eco-friendly" without proof
- Claim "biodegradable" for a compound containing silicone
- Use "natural" to describe a chemically modified derivative without context

Golden Rule:

"If you can't measure it, don't claim it."

Final Thought: Sustainability Is a Performance Metric

You wouldn't accept a dispersant that failed to reduce specks.
Don't accept one that fails to reduce emissions, contamination, or waste.

Sustainability is not a separate department.
It's a design criterion, as critical as color strength, impact resistance, or die life.

The most sustainable compound isn't the one with the least additive.
It's the one that:

- Performs reliably
- Uses renewable or recyclable inputs

- Doesn't contaminate recycling streams
- Leaves no toxic trace

Choose your dispersant like you choose your safety gear:
Not because it's cheap.
Not because it's easy.
But because it's the right tool for a responsible future.

21. Future Trends and Innovations

The future of dispersion in plastic processing is not about incremental improvements, it's about system-level transformation. As sustainability mandates tighten, performance expectations rise, and digitalization accelerates, the next generation of dispersing and wetting additives will be smarter, more precise, and deeply integrated into the entire value chain. Below are the four most impactful trends shaping the future, each grounded in current R&D, pilot trials, and market signals.

21.1 Nanodispersions and Hybrid Dispersants

Nanodispersions and hybrid dispersants aren't lab curiosities, they're the next step for formulators who've hit the limits of conventional additives. When you're working with nano-silica, nano-TiO_2, or surface-treated clays, standard waxes and EBS no longer cut it: the particles clump too fast, too hard, and too small to see until they wreck your film clarity or die stability. Nanodispersions use purpose-built block copolymers or grafted brushes to lock individual nanoparticles in place, not just wet them, but shield them. Hybrids go further: a single molecule that couples and lubricates, or stabilizes and slips, cutting additive count, cost, and compatibility risk. This isn't about chasing novelty. It's about solving real problems, like UV haze in PET bottles or die gumming in high-nano-fill TPEs, without adding more variables to an already overloaded formulation.

The next frontier is not just *smaller* particles, but *engineered interfaces*.

Nanodispersions:

Moving beyond micron-sized pigments and fillers, formulators are now working with nanoparticles (1–100 nm): nano-silica for barrier films, nano-clay for flame retardancy, and nano-TiO_2 for UV protection. But nanoparticles have extreme surface energy, they agglomerate violently unless perfectly wetted.
→ Solution: *Nanodispersion systems* using block copolymer surfactants or

grafted polymer brushes that form dense, steric shields around individual nanoparticles. These are not traditional dispersants, they're nanoscale encapsulants.

Example: A German compounder using a custom PEO-PPO-PEO triblock copolymer to stabilize 20 nm TiO_2 in PET for food packaging. Result: 50% reduction in UV haze, no migration, and full recyclability.

Hybrid Dispersants:

The future lies in multi-functional molecules that combine roles previously handled by separate additives.

→ Examples:

- A maleated polyolefin grafted with a silicone segment, couples filler *and* reduces die build-up.
- A phosphate ester with a bio-based wax tail, stabilizes PLA *and* provides slip without blooming.
  → Benefit: Fewer additives = simpler formulations, lower cost, reduced regulatory burden, and fewer compatibility risks.

Industry Prediction: By 2030, 40% of new masterbatches will use hybrid dispersants instead of traditional blends.

21.2 Smart Additives (Self-Adjusting or Reactive Dispersants)

Smart additives aren't sci-fi, they're chemistry that responds instead of just reacting. Think of them as molecular gatekeepers: inactive during storage or low-shear zones, but switching on when they sense the right temperature, shear, or moisture level to wet, stabilize, or protect. A thermo-responsive dispersant, for example, stays folded and non-migrating at room temperature but unfolds at 190°C to grab carbon black and lock it in place, then retracts on cooling to prevent plate-out or bloom. These systems cut additive count, reduce trial-and-error, and solve problems that fixed-chemistry additives can't, like migration in sterilized medical parts or silica reactivity in TPEs.

This isn't about complexity for its own sake. It's about additives that work harder, not just more.

Additives that *react* to their environment, this is no longer science fiction.

Thermo-Responsive Dispersants:

Polymers that change conformation with temperature.
→ At low temps: compact, inert.
→ At processing temps (>180°C): unfold, expose polar groups to bind pigment/filler.
→ At cooling: retract, minimizing migration.

Application: Masterbatches for injection molding where surface gloss must be high *after* cooling, but flow must be excellent *during* extrusion. No bloom, no plate-out.

pH- or Moisture-Responsive Systems:

Designed for biodegradable or compostable systems.

→ In PLA: A dispersant that remains inactive during processing but activates upon composting to accelerate filler degradation.
→ In water-exposed applications: A dispersant that releases a protective agent only when humidity exceeds 70% (e.g., for outdoor cable insulation).

Pro Tip: These are not "magic bullets." They require precise molecular design and rigorous validation. But when they work, they eliminate 2–3 traditional additives.

Self-Healing Interfaces:

Early-stage research into additives that can reform broken polymer-filler bonds under stress or heat.
→ Inspired by biological systems (e.g., mussel adhesive proteins).
→ Could revolutionize recycled composites, where interfacial degradation is the #1 cause of performance loss.

Bottom Line:
"The next generation of dispersants won't just help you mix, they'll help you adapt."

21.3 AI-Assisted Formulation Design

AI-assisted formulation isn't about replacing compounders; it's about cutting the guesswork out of trial-and-error. Think of it as a smart lab partner that's already run thousands of virtual batches: it knows that EBS spikes filter pressure in silica-filled TPEs, that 1.2% BYK-P 4100 hits $\Delta E < 1.0$ in black PP, and that 0.8% PE-g-MA won't cut it for 45% $CaCO_3$ in HDPE pipe. You feed it your resin, filler, target specs, and processing limits, and it spits out a shortlist of additive options, dosages, and even extruder profiles worth testing. No hype, no black boxes, just data-backed starting points that get you to first-pass success faster. In today's world, where a week of R&D downtime costs more than an AI license, it's not magic. It's math you can trust.

The era of "trial-and-error" compounding is ending.

AI Platforms in Action:

Companies like BASF, Covestro, and start-ups like ChemIntel are deploying machine learning models trained on decades of formulation data:

- Polymer type
- Additive chemistry
- Processing parameters
- KPIs (MFI, ΔE, impact, filter pressure)
- Regulatory status
  → The AI predicts optimal additive type, dosage, and blend for a given performance target, *in seconds*.

Real-World Impact:

A U.S. compounder needed a black PP masterbatch with ΔE < 0.8, MFI > 14, and zero plate-out for a high-gloss automotive part.

- Traditional method: 12 lab batches over 6 weeks.
- AI model: Recommended 1.1% BYK-P 4100 + 0.2% Struktol TPW 150.
- Result: Achieved target on first trial. Saved $85,000 in R&D cost.
- AI also flags hidden incompatibilities, e.g., "Using EBS with this silica grade increases die build-up risk by 78%."
- Integration with Digital Twins:
  AI models are now linked to digital twins of extruders.
  → Simulate how a new formulation behaves under different screw speeds, temperatures, and residence times, before running a single kilo.
  → Predicts optimal processing profile *and* additive package simultaneously.

Future Outlook:
"In 5 years, your formulation scientist won't be flipping through MSDS sheets, they'll be training an AI model and validating its predictions on the lab extruder."

21.4 Regional Market Trends (EU, Asia, MENA)

Dispersion isn't one-size-fits-all anymore, what works in Germany fails in Guangzhou and gets rejected at a Saudi port. In the EU, it's regulation driving change: REACH, CBAM, and the Packaging Directive are phasing out EBS, silicones, and anything that gums up recycling streams. In Asia, it's scale and cost: local PE-g-MA and bio-wax blends are displacing imports, and AI-driven QC is standard in large masterbatch plants. In MENA, it's diversification: bio-based additives from non-food crops and solar-powered compounding are positioning the region as a green export hub for EU markets. This isn't about global best practices. It's about matching chemistry to geography, because the right dispersant in one market can be a liability in another.

Dispersion innovation is no longer global, it's geographically fragmented by regulation, cost, and raw material access.

Table 21.4: Regional Market Trends (EU, Asia, MENA)

Region	Key Driver	Emerging Trend	Strategic Implication
European Union	Regulation-Driven (REACH, CBAM, SUPD, Packaging & Packaging Waste Directive)	•100% bio-based, compostable dispersants for food packaging •Zero-tolerance for silicone in recycling streams •Carbon footprint labeling of additives	• Invest in ISCC PLUS-certified raw materials • Phase out EBS and low-MW waxes in food-contact applications • Partner with recyclers to co-develop "recyclable-compatible" formulations
Asia (China, India, SE Asia)	Cost & Scale-Driven (Mass production, low labor cost, rising domestic demand)	•Localized, low-cost alternatives to imported dispersants (e.g., Chinese-made PE-g-MA) •Hybrid formulations using 50% recycled content + bio-wax • Rapid adoption ofAI-driven QC in	• Source locally to cut cost and lead time • Don't assume "universal" additives work, regional resins (e.g., Indian PP) behave differently • Focus on high-volume, mid-

		large masterbatch plants	performance solutions
MENA (Middle East & North Africa)	Diversification & Sustainability-Driven (Oil-to-chemical transition, green hydrogen, Vision 2030)	•Bio-based dispersants from local crops (e.g., camelina, jatropha) •Solar-powered compounding with low-energy additives • Export-focused premium packaging for EU markets	• Build partnerships with regional agro-industries • Position your formulations as "green enablers" for export • Invest in digital trace ability to meet EU import requirements

Critical Insight:

"The 'global standard' is dead. The future belongs to regional champions who understand local feedstocks, regulations, and customer pain points."

Final Outlook: The Next Decade of Dispersion

The future of dispersing and wetting additives is defined by three converging forces:

1. Precision, from "good enough" to "perfectly engineered" at the molecular level.
2. Intelligence, From human intuition to AI-guided, data-driven formulation.
3. Responsibility, From compliance to leadership in circularity and decarbonization.

The companies that thrive will not be the ones with the cheapest additives. They will be the ones who:

- Design for the entire lifecycle, from feedstock to landfill or reprocessing,
- Leverage data to eliminate waste, in R&D, production, and quality control,
- Speak the language of regulators and recyclers, not just chemists.

This isn't a shift in chemistry.
It's a shift in mindset.

Your next formulation won't just make a better part.
It will make a better planet, and a better business.

22. Appendices

22.1 Appendix A – Glossary of Terms

Table 22.1: Key Glossary of Terms

Term	Definition
Agglomeration	The clumping together of fine particles (e.g., pigment, filler) into larger clusters due to interparticle forces. Agglomerates are not dissolved; they are physically stuck together.
Dispersion	The process of breaking down agglomerates into individual primary particles and distributing them uniformly throughout the polymer matrix.
Wetting	The initial step where the polymer melt replaces air or moisture on the surface of a solid additive (e.g., pigment, filler), allowing intimate contact.
Stabilization	The mechanism that prevents re-agglomeration after dispersion by creating a physical or chemical barrier around particles.
Plate-out	The migration and deposition of low-molecular-weight additives (e.g., EBS, PE wax) onto metal surfaces (screws, dies, molds), causing buildup and surface defects.
Bloom	The migration of an additive (e.g., stearate, erucamide) to the surface of a finished part over time, often appearing as a white, hazy, or oily film.
Melt Flow Index (MFI)	A measure of polymer melt viscosity under standardized temperature and load. Used to monitor

	flow changes caused by dispersion, degradation, or additive loading.
Delta-E (ΔE*ab)	A numerical value representing the total color difference between two samples. $\Delta E > 1.5$ is typically visible to the human eye.
Haze	The percentage of transmitted light that is scattered, causing cloudiness in transparent or semi-transparent plastics. Caused by poor dispersion, additive migration, or moisture.
Gloss (GU)	Gloss Units, a measure of specular reflectance at a defined angle (e.g., 60°). Higher values indicate smoother, shinier surfaces.
Coupling Agent	A chemical additive (e.g., PE-g-MA, PA-g-MA) that forms a chemical bridge between a polar filler/fiber and a non-polar polymer matrix, improving adhesion and mechanical properties.
Compatibilizer	A type of coupling agent, often a grafted polymer (e.g., SEBS-g-MA), used to improve interfacial adhesion and reduce phase separation in polymer blends or composites.
Steric Stabilization	A stabilization mechanism where long polymer chains extend from the additive surface into the melt, physically preventing particles from approaching each other.
Electrostatic Stabilization	A mechanism common in liquids (e.g., paints) where charged particles repel each other. *Not effective in non-conductive thermoplastics.*

Residence Time	The total time a polymer compound spends in the molten state during extrusion or injection molding. Critical for additive activation and dispersion.
Shear Rate	The rate at which adjacent layers of molten polymer move past each other. High shear breaks agglomerates; excessive shear can degrade polymer or break fibers.
Filter Pressure Rise	The increase in pressure across a screen pack during extrusion. A rapid rise indicates agglomerates are clogging the screen, a direct indicator of poor dispersion.
REACH	EU Regulation (EC) No 1907/2006 on Registration, Evaluation, Authorization and Restriction of Chemicals. Requires full disclosure of Substances of Very High Concern (SVHC).
RoHS	Restriction of Hazardous Substances Directive (EU 2011/65/EU). Restricts lead, mercury, cadmium, and other hazardous materials in electrical/electronic equipment.
FDA 21 CFR 178.3820	U.S. regulation governing indirect food additives, including lubricants and dispersants that may migrate into food.
EN 13432 / ASTM D6400	International standards for industrial compostability of packaging materials. Requires >90% biodegradation within 12 weeks.
RSPO	Roundtable on Sustainable Palm Oil. Certification ensuring palm oil derivatives (e.g., ESBO, GMS) are sourced without deforestation or human rights violations.

ISCC PLUS	International Sustainability & Carbon Certification. A globally recognized standard for sustainable and traceable bio-based and recycled materials.
Log (Letter of Guarantee)	A formal, signed document from a supplier confirming a product's compliance with regulatory standards (e.g., FDA, REACH, food contact).

Training Tip: Use this glossary in onboarding sessions. Ask new technicians to define terms before and after training, measurable improvement = better communication.

22.2 Appendix B – Conversion Tables

Practical conversions for global teams and equipment calibration.

Table 22.2: Key Conversions

Quantity	Commission
Temperature	°C → °F:°F = (°C × 1.8) + 32 °F → °C:°C = (°F – 32) ÷ 1.8
Concentration	% by weight → Parts Per Hundred (phr):1% = 1 phr phr → %:1 phr = 1%
Pressure	Bar → psi:1 bar ≈ 14.5 psi psi → bar:1 psi ≈ 0.069 bar
Melt Flow Index (MFI)	g/10min → g/5min:Multiply by 0.5 g/10min → g/2min:Multiply by 0.2
Film Thickness	Microns (μm) → mils:1 mil = 25.4 μm μm → mils:1 μm ≈ 0.0394 mil
Viscosity	Pa·s → cP (centipoise):1 Pa·s = 1000 cP

Length	mm → inches:1 inch = 25.4 mm mm → inches:1 mm ≈ 0.0394 inches

Pro Tip: Print and laminate this table for the extrusion line control room. Avoid calculator errors during shift changes.

22.3 Appendix C – Quick Reference

Table 22.3.1: Dispersant Selection by Polymer Type

Polymer	**Recommended Dispersant**	**Avoid/Use with Caution**
PE, PP, PS	PE wax, PP wax, Montan wax, Silicone masterbatch	Surfactants (PEG esters), EBS (if high gloss required)
PVC	Calcium stearate, EBS (low dose), Epoxidized soybean oil	Silicone, Zinc stearate, Metal soaps (with Pb/Cd)
PA (Nylon)	PA-g-MA, SEBS-g-MA	PE wax, EBS, Paraffin wax
PBT, PET	PET-g-MA, PA-g-MA, Phosphate esters	EBS, Low-MW waxes
PC, ABS	Struktol® TPW, BYK-P 4100	Silicone (if painted), EBS
PLA	Glycerol Monostearate (GMS), Epoxidized Soybean Oil (ESBO), Bio-wax	PE wax, EBS, Silicone
TPE (SEBS/EPDM)	Silastic™ MB-100, Titanate coupling agents	EBS, Erucamide (in silica-filled systems)

Recycled PE/PP	PP-g-MA + PE wax + Stearic Acid	Silicone, High-MW waxes

Use this chart as a first-pass decision tool. Always validate with KPIs.

Table 22.3.2: Additive Incompatibility Matrix

Additive	Incompatible With	Risk/Consequence
EBS	Silicone masterbatches, PA, PET, High-temp systems (>240°C)	Migrates, blooms, degrades, causes odor
Silicone	Paints, inks, adhesives, EBS, PA, PVC	Causes fish-eyes, reduces adhesion, contaminates recycling
Calcium Stearate	Zinc stearate, Peroxide-cured systems	Forms zinc chloride → catalyzes degradation
EVA	High-temp systems (>205°C), PA, PET	Degrades, releases acetic acid
Phosphate Ester	Highly polar resins (e.g., PC)	May reduce impact strength if overdosed
PE-g-MA	High-MFI virgin PE (in recycled blends)	Dilutes recycled content; may reduce compatibility

Print and post in the formulation lab. Rule: "If unsure, test before you scale."

Table 22.3.3: KPI Target Thresholds by Application

Application	KPI	Target (Pass)	Failure Threshold

Color Masterbatch (PP)	Speck Count	≤ 2 specks / 10 cm²	> 5 specks / 10 cm²
	ΔE (after aging)	≤ 1.2	> 2.0
	MFI Change	±0.5 g/10min from base	> ±1.5 g/10min
Filler Masterbatch (HDPE + $CaCO_3$)	Filter Pressure Rise	≤ 5 bar / 2h	> 10 bar / 2h
	Tensile Modulus	≥ 2000 MPa	< 1800 MPa
	Haze	≤ 5%	> 8%
Engineering Plastic (PBT + Glass Fiber)	Fiber Length Retention	> 85%	< 80%
	HDT @ 1.82 MPa	≥ 230°C	< 220°C
	Die Build-Up	> 48 hrs between cleaning	< 12 hrs
Food-Contact Film (LDPE)	Migration (FDA 178.3820)	< 0.05 ppm	> 0.1 ppm
	Seal Strength	≥ 1.5 N/mm	< 1.0 N/mm
	COF (Dynamic)	≤ 0.2	> 0.4
Recycled PE + $CaCO_3$	Tensile Strength	≥ 12 MPa	< 10 MPa
	Filter Pressure Rise	≤ 8 bar / 2h	> 12 bar / 2h
	Visual Specks	≤ 3 / 10 cm²	> 5 / 10 cm²

Medical Device (PP + TiO_2)	ΔE (post-sterilization)	≤ 1.0	> 1.5
	Whiteness Index	≥ 88	< 85
	Sterilization Pass Rate	100%	< 95%

Integrate these thresholds into your SOPs and quality control checklists.

22.4 Appendix D – Supplier Reference List with verified compliance info

This table is a living document. Update quarterly. Always request and retain current Letters of Guarantee (LOG).

Table 22.4: Example Supplier Reference List with verified compliance info

Trade Name	**Supplier**	**Primary Function**	**Key Compliance Certifications**	**Notes**
Licowax® PE 520	Clariant	PE Wax / Lubricant	FDA 21 CFR 178.3820, REACH, RoHS	Gold standard for HDPE/PP. Low bloom.
Hydrocerol®	Clariant	Wax-based Dispersant	FDA, EU 10/2011, ISCC PLUS	Ideal for food-contact color masterbatches. Non-migrating.
Struktol® TPW 100/150	Schill + Seilacher	Multi-functional Dispersant	REACH, RoHS, ISO 9001	Excellent for BFR, carbon

				black, high-temp systems.
BYK-P 4100	BYK-Chemie	Polar Dispersant	FDA, REACH, RoHS	Best-in-class for carbon black, TiO_2 in PP/PA.
Silastic™ MB-100	Dow	Silicone Masterbatch	FDA, REACH, RoHS	For silica-filled TPEs. Low volatility.
EBS (Ethylene Bis-Stearamide)	Generic / Multiple	Lubricant / Dispersant	Varies by supplier	*Always request LOG.* Avoid unbranded sources.
Montan Wax	Montan Wax GmbH	Natural Wax	FDA, REACH	High thermal stability. Preferred in rigid PVC.
Glycerol Monostearate (GMS)	Evonik / Kao	Bio-based Wetting Agent	FDA, EU 10/2011, ISCC PLUS	Essential for PLA and bio-based systems.
Epoxidized Soybean Oil (ESBO)	Archer Daniels Midland	Bio-based Plasticizer	FDA, EU 10/2011, RSPO	Critical for PVC and PLA stabilization.
PP-g-MA (e.g., A-300)	Arkema / Lanxess	Coupling Agent	REACH, RoHS	Use only virgin-grade for food contact.

Tyzor® TPT	Dow	Titanate Coupling Agent	REACH, RoHS	Essential for silica, clay, and WPCs.
Irganox® 1010	BASF	Antioxidant	FDA, REACH	Prevents oxidation in PE, PP, PS, and PLA.

Compliance Protocol:

1. Before purchase: Request current LOG for *your specific application* (e.g., "FDA for food-contact film").
2. Upon receipt: Verify batch number matches LOG.
3. In storage: File LOG with batch record.
4. Annually: Audit supplier compliance status.
5. Never rely on generic SDS or website claims.

This appendix is your legal shield. Maintain it like your quality manual.

22.5 Index of Tables

www.ingramcontent.com/pod-product-compliance
Ingram Content Group UK Ltd.
Pitfield, Milton Keynes, MK11 3LW, UK
UKHW051509290926
15471UKWH00008B/1521

9 798232 768409